Global Challenges
for Leviathan

Global Challenges
for Leviathan

A Political Philosophy of Nuclear
Weapons and Global Warming

Furio Cerutti

LEXINGTON BOOKS

A division of
ROWMAN & LITTLEFIELD PUBLISHERS, INC.
Lanham • Boulder • New York • Toronto • Plymouth, UK

LEXINGTON BOOKS

A division of Rowman & Littlefield Publishers, Inc.
A wholly owned subsidiary of The Rowman & Littlefield Publishing Group, Inc.
4501 Forbes Boulevard, Suite 200
Lanham, MD 20706

Estover Road
Plymouth PL6 7PY
United Kingdom

British Library Cataloguing in Publication Information Available

Library of Congress Cataloging-in-Publication Data

Cerutti, Furio, 1938-
  Global challenges for leviathan : a political philosophy of nuclear weapons and global
warming / Furio Cerutti.
     p. cm.
  Includes bibliographical references and index.
  ISBN-13: 978-0-7391-1687-6 (cloth : alk. paper)
  ISBN-10: 0-7391-1687-8 (cloth : alk. paper)
  1. Nuclear weapons. 2. Nuclear arms control. 3. Global warming. I. Title.
  JZ5665.C47 2007
  320.01-dc22

                                                          2007022701

Printed in the United States of America

∞™ The paper used in this publication meets the minimum requirements of American
National Standard for Information Sciences—Permanence of Paper for Printed Library
Materials, ANSI/NISO Z39.48–1992.

Contents

Preface

This book is a philosophical reflection upon the end of modernity, and in particular of the relatively rational political order that was established in Europe and later the Americas beginning with the Peace of Westphalia of 1648, was shattered in the previous century by the world wars, including the Cold War, and is now further challenged by planetary threats to our environment.

This reflection is however not conducted, as is usual in philosophical books, by rerunning through other philosophers' doctrines and suggesting more or less explicitly that reinterpreting them gives us the ability to illuminate the present situation. In my view, our era (this, not a span of a mere few years, is the matter of philosophical reflection) is characterized by epoch-making facts that have been given hardly any attention and undergone little elaboration in philosophy. More than the set of phenomena and processes that we know by the title of globalization, it is the twin threats represented by the sheer existence of nuclear weapons and by global warming that endanger human civilization, rather than the biological existence of the human race. This now happens to an extent that justifies thinking of all this in terms of survival problems; their philosophical relevance goes far beyond the importance of currently much debated issues such as terrorism or multiculturalism. Why I call them global challenges, and why I presently recognize only those twin threats as global challenges is explained in the Introduction. More radically than any other factor they show the inability of Leviathan, the modern state, to come to terms with the dangerous sub-effects of modern science and technology and to still provide effective protection for the citizens and reasonable management of the global commons with an eye to future generations. Democracy in its meaning as rule-based collective decision-making over our fate is also threatened by this fate now being dependent on events that human beings have set in motion, but are unable to govern.

Philosophers sometimes have an elusive relationship with the facts that define the state of the world they are reasoning about; facts and trends are evoked in an impressionistic style rather than analyzed in their logics. I have tried to steer clear of this danger by analyzing the two global challenges at some length, and this explains the interdisciplinary nature of this book. Political science, in particular International Relations, is the other main discipline in my methodology, with some contribution from general social theory and international law as well as modern, including nuclear, history. As the threats examined are first of all physical processes, the reader will also find some basic information from physics and especially climatology. Political philosophy plays the leading role in coordinating the various insights and discoveries of new problems made by the other disciplines, in finding names or rather categories for them and putting them in an overarching argument: global challenges give the final blow to modern politics and raise the question of an adequate change in our mindset and of new institutions capable of addressing them, granted that it is not too late. As a result, this book is neither bent on fostering catastrophism (the challenges can still be taken up, and must be continuously reassessed) nor on joining postmodernism (we are at the end of modernity, but we cannot survive without redesigning the polity and reshaping rationality).

The reader may have noted that this political philosophy is not identical with the usual, strictly normative meaning of the term, particularly in the English-speaking world. I am indeed interested in analyzing and theorizing the state of the world and in giving philosophical names to the resulting problems rather than concentrating on formulating a reasoning about what is just, which does not take note of what is possible and how the actors can be expected to act. The ambition of this book is to go beyond the traditional dispute of political realism vs. normativism; in doing so it obviously discusses present and previous doctrines whenever this can help achieve a better understanding of the substantive problems that are the book's main object, but it cannot go into all the intricacies of the theories and authors cited.

My theoretical description of what is going on at the end of modernity then goes over into normative questions: should we do something to address global challenges, and why? Should we only take care of ourselves, or the coming generations as well? Should we put together a new Superleviathan to protect ourselves or are there other possible solutions? As can be expected from philosophy's critical approach, no answer is presented as obvious, and counterarguments are taken into serious consideration. Those not convinced that global challenges are a problem that must be actively addressed can find their positions duly debated, those taking the opposite position can find their reasons clarified and strengthened by a detailed argumentative foundation.

Both parties will find more problems raised and arguments discussed than recipes provided.

This work has been written over a long span of time, not only because the multiplicity of its topics required a great deal of research, but also because for a long while I remained uncertain as to the institutions that can come to terms with global challenges, particularly with regard to the nuclear arms issue. The growing importance of man-made climate change and the literature dealing with this issue helped me regain the thread that finally led me to complete the book. This gives me the welcome opportunity to bring back to theoretical attention, so I hope, the nuclear weapons issue (rather than nuclear proliferation, which is only a corollary), an issue that philosophers as well as political scientists too easily laid aside after the end of the Cold War. The impulse to deal with nuclear weapons and what they mean for humankind's fate on earth first came to my mind over twenty years ago, during the last nuclear arms race of the Cold War, which prompted me not only to be reasonably scared, but also to take on the philosophical exercise of peering into the origin and meaning of those tools of annihilation.

Acknowledgments

This book was conceived and to a large extent written during my repeated stays at Harvard University, first at the Law School and later at the Minda de Gunzburg Center for European Studies. I am deeply indebted to Harvard's magnificent tradition of hospitality and to all those who made my stays possible and productive. In particular, I remember a clarifying conversation with Bob Keohane, then still at Harvard, on universalism and realism.

I also wish to express my grateful memory of scholars who, now no longer among us, gave me encouragement and valuable hints on problems that were later explored in this book: Norberto Bobbio at Turin, whose writings first introduced me to the political philosophy of international relations, Abram Chayes of Harvard Law School, Ernst Haas at Berkeley, and, not far from Cambridge, Viktor Weisskopf, one of the European physicists who worked at Los Alamos on the atomic bomb project.

Many years ago my first readers were Michael Brenner and Rodolfo Ragionieri; they gave me the impulse to further venture onto this quite complicated path of research. Later, Sonia Lucarelli gave me not only encouragement, but also an essential helping hand in connecting to state-of-the-art international relations. She discussed the entire manuscript with me, as did Dimitri D'Andrea, while Elena Pulcini and Chiara Bottici commented on several chapters. My deepest thanks to all of them as well as to the two physicists, Bruno Carli and Franco Strumia, who helped me rid the text of some imprecisions.

I had the opportunity to discuss parts of this book at the Université de Paris 8, the Humboldt as well as the Free University in Berlin, the Universities of Siena and Pavia, the Scuola Superiore S.Anna in Pisa and the Società Italiana di Filosofia Politica in Rome. On these occasions many colleagues, among them Gérard Mairet, Eleni Varikas, Volker Gerhardt, Ian Carter and others, whose helpful criticism is quoted alongside the passages in question, gave me valuable comments for which I am thankful. My thanks also go to the anonymous reviewer who read the complete version of this work.

In preparing the final version I took advantage of the research assistantship provided by Elena Acuti, while Karen Whittle made my English more acceptable to native speakers; I am indebted to the both of them, while I remain responsible for the present shape of the book.

A less aggressive approach to nature is a theme of this book, and this also finds expression in my gratitude to the living beings, the squirrels in Harvard Yard and the cats at our home in Tuscany, who made me smile on days in which the burden of researching and writing and the sadness of topics such as nuclear annihilation inclined me to rather a somber mode.

The patience and encouragement of my wife, Renata Carloni, accompanied this work from its very inception, but this is only one of the reasons why I dedicate it to her.

Introduction

In this introduction I am trying to explain what I mean by "global challenge," why we can presently recognize only two of them, one of which, the nuclear weapons issue, is recalled here in its basic elements and also with an eye on the overstated question of an extinction of humankind. I shall finally make clear that global challenges are not a sub-chapter of globalization, while I am not going to discuss at length the methodology I have developed throughout the chapters. I hope its reasons are sufficiently clear in the several passages in which my philosopher's excursions into multi- or interdisciplinarity, primarily international relations, are justified with regard to the specific matter under consideration.

1. DEFINITIONS, DELIMITATIONS, AND A ROAD MAP

What kind of book is this?

In the intention of its author, it is a primarily, though not exclusively philosophical book, involving political philosophy, ethics, and the philosophy of civilization. Its initial subject matter, nuclear weapons and global warming, is however not made of issues that everybody would say are indisputably philosophical; nuclear weapons and global warming are rather referring to physics and climatology and/or political science. My not so hidden assumption is that those issues tell us more about the human condition now and in the future than explicit and current philosophical concepts and debates. But this assumption must be painstakingly argued all throughout the chapters and does not pretend to be shared intuitively by the reader. The long journey to the philosophical core, which at the end will take us to the sphere of politics, sets out therefore in this introduction with the attempt to demarcate this book's topics.

Unlike most of the critique of civilization produced by both conservative and radical philosophers in the first half of the last century, this book will abstain from giving a general and impressionistic picture of the catastrophes and evils to be. The planetary threats that we have ourselves created and under which we live may lead humankind to a tragic end, but it is not hopelessly illusory to think that we still have a chance to contain them; this is why it makes sense to think of them in terms of challenges to our ability to survive (by reshaping our mind and reorganizing the polity) rather than of mere threats. A first step in this direction is to give a narrow *definition of global challenge*,[1] which consists of two parts.

Global challenges are, first, only those that, whatever their roots in social and political processes, consist of (actual or expected) physical phenomena which can be described, measured, and disputed by means of experimental science. To be sure, even in this kind of science, facts are theory-dependent, as is well known to everybody who acknowledges the probabilistic and fallibilistic status of scientific knowledge. Nonetheless, our expectation of facts like the further depletion by a factor of so-and-so-much of the ozone layer, if the emission of chlorofluorocarbons is not effectively reduced by a corresponding factor, is in a point of reliability quite different from the prediction of, say, the definitive commodification of human relations, or, in the field of globalization studies, the "death of the distance." As of now, science provides reliable accounts of the challenges that may endanger our future for just some of the many issues drawing our attention. Prophetic accounts of the future, which combine all of them, are stimulating, can be useful, and are in some case, of esthetic interest; but this is intended to be a plainly scholarly, not prophetic book. It will therefore discuss only the two challenges for which solid evidence can be provided, even running the risk of setting aside trends (say, the political effects of biotechnology) whose perhaps greater relevance will come up in future times.

In the second leg of the definition, the use of the adjective "global" is reserved for those physical phenomena that can directly affect humankind as a whole, both as object and subject (or as target and actor), in a nearly equal and potentially lethal manner. As a whole: not only can all men and women who happen to live at a given time be affected, but the civilized existence itself of our kind is possibly at stake. As object as well as subject: while everybody is concerned or threatened, the feared outcome can be averted only by the cooperation of all, or at least the overwhelming majority of both the individual inhabitants of this planet and their polities. In a nearly equal manner: for a threat to be truly global, everybody must be more or less severely hit by the related phenomena, and safe havens must be inaccessible for whatever (geographical, social, or ethnic) group of relevant size. Potentially lethal or, in other words, "ultimate challenges": as they affect, though in different degrees, the survival of our kind and civilization.[2]

There are two philosophical reasons for the pre-eminence of global issues over other issues of greater actual concern (for instance problems of international distributive justice, which do not match the standards I have given for globality) and/or visibility, such as international terrorism, which, besides not being global, is likely to be overstated if seen in a larger historical perspective.[3] For moral philosophy humankind, not a group or other intermediate levels of agency, is the sole notion whose relevance nears that of its classical sphere, the individual morality; humankind is indeed likely to become moral philosophy's second focus and to emerge as crucial in political philosophy as well, also in connection with the problems raised by economic and social globalization. The second reason regards most specifically political philosophy: only global, ultimate, and generally acknowledged threats can generate forces driving the divided and conflicting humankind into a community not based on goodwill, universal love, or argumentative persuasion, which all fall short of creating a political community. At the outset of modernity Thomas Hobbes taught us that only ultimate threats can push the recalcitrant individuals into the process of establishing the polity, and we are now asking if and how far the present global challenges can have the same effect. Not very far, as we shall see, as "can generate forces" means nothing more than a chance, which can be spoiled by counterforces or unexpected processes. Nor would the political community resulting from the pressure of global challenges be necessarily like the world government of so many utopian projects. But for political philosophy the sole glimpse of that chance makes global challenges a subject worth closer examination, and this is the main justification for writing a book upon them. How will it develop?

After this Introduction, the book's road map starts in Part One with the substantive description of the two challenges, starting with the establishment and later the upsetting of modern political order (chapters two and three) and ending with global warming (chapter four). Chapter one is reserved for the preliminary task of preserving the two global challenges from being submerged in the vast sea of "risk" and "risk society," thus losing their philosophical stringency. A substantive description means recalling the basic physical data as well as the historical developments that have made certain technological side-products of our scientific and economic advancements a challenge for Leviathan, the modern polity. In Part Two I then examine how far various philosophical concepts and doctrines (normative ethics in chapter five, but also the philosophy of meaning in chapter six) can contribute to a practical attitude towards global challenges and try to develop my own approach, which is also concerned with the role of science and technology in modernity. In chapter seven I finally ask what this means for a politics that, coming after modernity, may want to live up to those challenges. Following the intention I have mentioned in the Preface, I have started in Part One with a substantive approach to them rather than exhibiting at the outset a philosophical system, which I do not in any case possess, and in which the specific links to common experiences and challenges are lost or unaccounted for.[4] I ought also to warn the reader from looking for solid and vocal policy recommendations in the last, political chapter. In my cautious approach to political solutions there is less resignation before a world that rejects the precepts of Reason than a reinterpretation of the role of political philosophy, aimed at keeping it free from futile ambitions. Its dumping sites are sufficiently full of projects of eternal peace, world order, and salvation of humanity, which have however hardly contributed to give us a little more effective peace, justice, and confidence.[5]

2. A PROBLEMATIC CATALOGUE

Why does this book recognize only two issues as truly global, while excluding widely recognized and momentous problems such as overpopulation, famine, and AIDS epidemics, which conventional wisdom does also regard as global?[6] And what about bioengineering?

There are two types of reason for this: ontological and epistemological. In the first case issues are excluded because they are not global in the sense we have given to this notion; in the second we do not yet have enough theory and enough evidence to determine whether or not we are faced with global issues. Needless to say that it is not always easy or feasible to keep the two types of reason from intermingling.

It is easy to see that AIDS epidemics or the more recent SARS and avian flu are not global for ontological reasons, as they tend to spread out of control is only in regions with lower levels of wealth, education, and public health care. Far more complicated is the overpopulation problem: it has regional origins and spread, but it can affect the whole of humankind as far as it might cause shortage of food, water, and fuel, and sharpen the competition for their distribution up to the level of a "world civil war," while in any case worldwide pollution as well as greenhouse gases release would be greatly increased. There is no general agreement on demographic expansion being unsustainable in itself, or rather only under certain conditions (asymmetrical distribution of wealth and technological capacity); these conditions could be gradually and peacefully changed to an extent which, along with substantial innovation in food production, could allow to feed even an increased (beyond the present margin) number of inhabitants of our planet. As to the consequences for the environment, they could be perhaps averted by a general reduction of the *per capita* energy consumption and waste production, first and foremost in the affluent countries. I am personally convinced that the overpopulation problem cannot be completely solved if the birth rate, particularly in certain regions of the globe, is not brought to decrease, but the whole issue remains controversial and the different prospects are inevitably value-laden, as in the moral controversy over birth control. This is an epistemological reason, but even weightier is another, ontological one: the containment of overpopulation remains a national policy issue, as the case of China has shown. It has nothing to do with the compulsive necessity to apply altogether the same policy that denotes actions needed to address nuclear weapons or global warming.

The same can be said of famine, as far as it is related to overpopulation. On the one hand, famine seems to be unleashed in the first place by local circumstances such as the political breakdown of a country with subsequent disruption of its civil society. Other possible factors for the outbreak of famine are drought and desertification, a phenomenon with links to a truly global problem like global warming, which however still need to be fully ascertained; so far, global warming cannot be univocally seen as the unique or main cause, or as the triggering factor of those phenomena. In any case, for spectators writing or reading in the North of the world, it would be not just wrong, but hypocritical to regard as global threats famine or poverty, by which they and their children shall never be affected.[7]

The reasons I have resorted to in order to separate global from non-global challenges do not deny philosophical relevance to the latter: we can, for example, look at famine as a phenomenon which makes our identity and self-esteem as moral agents questionable, if we do not take action against it on a worldwide scale. To be true, how can we think of ourselves as bearers of universal norms, values, and rights, if we tolerate that, out of starvation and disease, other human beings die an inhumane death in the millions, whereas it would cost us just a little money and a serious organizational effort to avert this fate? But the very formulation of this question apparently involves assumptions about morality and economics that are far from being generally accepted and spark controversy, driving us to take lines. The "lifeboat Earth"[8] argument can be cited here; I shall not discuss its validity, but no doubt it must be taken seriously. What I am suggesting here is that issues of justice, such as those raised by famine and poverty, are structurally different from those of survival raised by the global threats, and not only because the latter put everything on earth in danger while the former do not. Issues of justice are about our idea of how men and women can best shape their lives in the human community, issues of survival[9] ask if there is going to be a human race living in community any longer, and what reasons we have for approving of humankind's survival as well as for acting in this sense. The substantive difference between the two classes of issues implies one of method: the survival issues raised by the global threats request parsimony in the assumptions about values, principles, and fundamental concepts, because an essential goal such as survival can be best secured by an essential argumentation, which does not require a lot of preliminarily shared beliefs and images-of-the-world. As I shall argue in chapter five, it is thin *versus* thick, the crude images of a Hobbesian world against the elaborate Hegelian pictures of a meaningful history of humanity. Not that I uphold a general realist view of the world; it is only that if the survival problems are not theoretically acknowledged and addressed, all that we can think and write about justice in society and among generations or a meaningful life looks like a building without ground floor, or something that lacks an essential component to make full sense. This book has been written, first of all, in order to touch on this blind spot in contemporary philosophy.

The exclusion of human biotechnology from this book's table of contents needs a more elaborate justification, in which epistemological reasons prevail, but are not the only to be relevant.

In the recent revolution unleashed by molecular genetics, and as far as the manipulation of human genetic material is concerned, it is still problematical to assess trends and threats. Five decades ago, shortly after the great scientific advancements of the fifties, some biologists prophesied the near possibility— after the huge increase in the quantity of human beings living on the earth— of improving their quality ("orthobiosis"): either by qualitative birth control or the genetically engineered creation of men with larger brains (Joshua Lederberg, Nobel Laureate of 1958) or without legs, replaced by a grasping tail, which would particularly fit the requirements of longtime living in a spaceship (J.B.S. Haldane).[10] In the last decades pictures like these have not been sketched with the same frequency and enthusiasm as in the sixties.[11] As far as we are allowed to know, no "Frankensteinian monsters" have been created. But it is difficult to assess the present state or the coming steps of gene manipulations such as the creation of reproductive clones and hybrids, using or not the chimera technology,[12] it is a question of both scientific-technical feasibility and direction of R&D under medical, economic, and ethical constraints. Even less predictable is what is going to happen with the normative regulation of the whole matter, at national as well as international (European Union, United Nations) level. But, apart from the unclear state of the question, why should the problems related to these developments be conceived of as global problems?

Let us first set aside direct military applications of biotechnology (so-called gene weapons, aimed for example at spoiling the enemy's crops). Let us also ignore the possibility of a cognitive misuse of genetic data for social or racial discrimination, which time ago ignited controversies about the Human Genome Project:[13] it would not differ from a Lombrosian social interpretation of anthropological data. What I am pointing at is rather a manipulative misuse of gene technology, carried out on a large scale and run by a political actor in order to strengthen its control over productivity and social uniformity and/or to enhance its might in interstate conflicts. These fears have been repeatedly evoked in the last century, before and after the discovery of DNA: most powerfully in Aldous Huxley's *Brave New World* of 1932 and in Ridley Scott's movie *Blade Runner* of 1982. The fact that they have so far not come closer to reality is no guarantee that they will never come into being: already the previous century has witnessed Stalin's Soviet Union failed attempt to obtain a "new invincible human being" by reproductive technology[14] as well as Hitler's *Lebensborn* project, which could theoretically be repeated on a less traditional basis, as soon as a totalitarian regime sweeps away the last restraints, a condition that as far as we know is presently not given. Nevertheless, this matter being relevant to moral and political philosophy, it is worth dwelling a little further on this mental experiment. From the political point of view, things would proceed in a manner like the nuclear arms race: once a country has decided to exploit the chances for supremacy it has been endowed with by an overwhelming technology, every other actor would be pushed towards developing and enhancing the same technique, be it focused on the prenatal selection of smarter and healthier individuals or the production of human clones. In this scenario, state-run eugenics or cloning would overrun individual will, claiming to be not a deprivation of personality and authenticity, but an enhancement of collective power and prestige, insofar a privilege for all participants.[15] Unlike in nuclear war and closer to man-made climate change not all members of humankind would be directly affected, but all would have to bear the consequences in terms of power relationships. What is more, our identity as human beings would be upset[16] and this very, if weak, cement of the human community would break apart. In a world in which the anthropological definition of "human" would be upset again and again in a willful, instrumental (to power gains), and abrupt way, not at the pace of the natural evolution from primate to human, how could we still recognize ourselves and our neighbors as those beings we are culturally and morally accustomed to think of, and to live with, as human?

It is true that at another historical turning point, the shift from the Middle Ages to modernity, we have been able to change the image of ourselves and to adapt to the new world designed by the experimental science of nature, a world of which man and earth were no longer at the center. But the creation of a Brave New World would be something completely different from redesigning the image of the old one: our ability to adapt to cultural change cannot be overstretched, as we have no guarantee that in every change we will be able to redefine, along with our identity, our dignity and richness, as it was the case in the shift to modernity. In other words: the maturity of both an individual and a society can certainly be seen as the ability to take chances and to face risks, but, while this ability has proven to be far greater than our forefathers may have thought, it is not unrestricted. Should the procreation of our offspring ever become subject to full and unrestricted manipulation, it is by far not said that adaptation to cultural change, to a different self-image, includes the ability for our posterity to still recognize themselves as members of one and the same kind, endowed with features that allow for continuity with the previous humanity. The challenge that would arise from this change would not mean an utterly extensive destruction of human beings and/or their environment and civilization, but could rather blow up the qualitative dimension of our being human; this is why it would be legitimate to include it among the global challenges. There would be logic in looking simultaneously at what has been discovered by the two leading sciences of the last century, physics and biology, in their own domains, matter and life.

Nonetheless this book will not explore the global challenge that may result from molecular biology because of the limited knowledge we have of some of its features, but also for two reasons, ontological and epistemological at the same time. First, how human bioengineering will develop is too uncertain to tell in terms of the possible success or failure. Second, we do not know how far it will be effectively and timely regulated by national and international norms that may prevent this branch of science and technology from developing out of any control and irrespective of unbearable consequences, as it happened with nuclear technology. The chance may not be high, but the outcome is too early to call. If reasonable regulation prevails, the developments that would make bioengineering a global challenge would simply not come into being. Third, at the time of writing there is no evidence of the political preconditions for such a challenge coming into being: democracies seem to be still able to keep the developments of biotechnology under control, and, while the planet does truly not lack authoritarian regimes, even among the great powers, hardly any important country seems to be on the verge of becoming totalitarian and even less of grasping the chances of the momentous gain in power that may lie in human bioengineering. Political science fiction can be an exciting literary or cinematic genre, but this book is written according to the standards of Western scholarship, which require a minimum of verifiable and arguable evidence in order to identify a research item. *Hypotheses non fingo*, to put it with Isaac Newton.

While the definition of what a global challenge is remains the job of political philosophy alone, the problem of which specific treats are to be seen as global challenges is an analytical one. Nuclear weapons and global warming have been chosen here as global challenges only because at this time there is enough evidence that they meet the criteria contained in the definition, while other candidate issues do not. But the set "global challenges" is an open set, potentially wider than the two items we presently find within it, and in future other items can be found to satisfy those criteria, or vice versa, new scientific evidence or (happily!) political reform can downgrade the global challenge status of the items now present in that set.

3. THE TWO PRESENT GLOBAL CHALLENGES

In this book nuclear weapons and global warming are the only threats that at this time I have reasons to acknowledge as global challenges. Here are the reasons.

Global warming is insofar a global issue as it is caused by the so-called greenhouse gases released by no matter what country, since all releases are well mixed in the atmosphere. It would be important if the worst polluters, the major industrialized countries, reduced significantly their emissions, as foreseen in the Kyoto Protocol; but a definitive solution cannot be achieved until all countries, including those which will later attain high levels of economic development, agree to cooperate now and, more importantly, in the future. On the other hand, not all of the predictable effects of the warming can be called global: moderate temperature increases may, *caeteris paribus*, inflict heavier tolls on poorer rather than wealthier countries and social classes, when it comes to adapting to the new climate. Northerners can take some advantage from the same higher temperatures that will harm Southerners; but both in the North and in the South low-lying areas may be lost to the increased sea level. But global warming as a whole, despite its differentiating effects, can escalate as a lethal problem, should its mechanisms get out of control, for example in the case that the Gulf Stream is diverted. Should this happen, the catastrophe that may result is not predictable with the same scientific reliability that seems to have been achieved in the case of nuclear winter; nor is it likely to be equally lethal, though remaining a catastrophe for our farther posterity. For these reasons, it makes sense to address global warming, or more exactly man-made climate change, as a global challenge with specific features; but it would be at least premature, or (hopefully) wrong to put global warming at the center of our philosophical reflections on ultimate challenges. We simply do not know enough to do so, nor has political science and philosophy yet generated as rich a reflection as on what in the 1950s received the nickname "nukes."[17]

The analytical part of this book is therefore largely (chapters two and three) devoted to nuclear weapons because they still represent the best-known paradigm of global challenge, not particularly in their present (first decade of the twenty-first century) reality, but rather as the permanent possibility of a major war. What this possibility means is well known from the history (for someone, the personal experience) of the Cold War, and the physical and strategic scenarios this book refers to are taken from it, because a fair deal of reliable research is available on that matter, and more importantly the balance of terror between West and East realized in full the potential of threat and fear contained in the weapons of mutual assured destruction. These scenarios may look aloof from the present situation, in which nuclear weapons are talked about only in the framework of proliferation and counter-proliferation efforts, sometimes with dramatizing accents.[18] No doubt a regional nuclear war (between India and Pakistan, say) or a local nuclear exchange in the framework of coercive counter-proliferation (a U.S. nuclear attack to disarm countries that are on the verge of going nuclear) would be not only horrible in themselves; they would also set the nuclear threshold at a lower level, or in the worst case ignite a major nuclear war involving great powers (say, China and Russia), unable to disentangle themselves. What I have in mind in the first place are however not these limited scenarios, but the possibility in a not so near future of a nuclear confrontation between existing nuclear powers with strategic arsenals, for instance USA and China. I am not predicting this, as I do not love resorting to commonplaces such as "history tells us that between the old and the new empire a supreme confrontation must take place." I only maintain that this scenario of doom cannot be ruled out politically and must be kept in mind philosophically.

A philosophically relevant corollary of this selection of global challenges must be underlined: they are both the result of human agency, consciously in the nuclear case, unwittingly (up to three decades ago) in the case of climate change. In other words, the threat is not caused by cosmic forces, as in the case of asteroids (cf. note 32 in § 5). These challenges thus suggest questions about the meaning of the human evolution on earth as well as guilt and responsibility; they will be addressed in chapters five and six.

Let us also anticipate that keeping that scenario of doom in mind and working on it means the rejection of two other choices. Instead of nurturing the technically impracticable and politically misleading dream of a world from which all nuclear weapons have been eliminated, presumably making the threat disappear, we should rather work in the direction of new institutions or institutional networks that can exert real and dependable control over them; but we should as well develop a corresponding political culture, based on a reasonable fear and a more balanced approach to the security problem. Second, I do not share the "utopian realism" (cf. chapter seven, § 3) of those who believe that every owner of nuclear weapons has learnt the lesson that they are useful just for deterrence, not at all for fighting a war, which will therefore never be fought.

On the contrary, the persistent availability of nuclear technology in the context of sovereign states, however engaged in international organizations, creates an open-ended situation of unrestricted threat. Even in the more peaceful international situation, nuclear technology cannot be disinvented, nor its attractivity for actors that may see a chance of power gains in it laid to rest. As mentioned above, neither the escalation of local nuclear conflicts into multilateral nuclear wars (with nuclear winter effects), nor the rebirth (probably with different actors) of actual tensions explicitly based on MAD[19] can be excluded. This statement is empirically not falsifiable, because nuclear weapons will be available in a future far ahead of us, which no scientific study of the present trends can predict as finally peaceful. It must however not be mixed up with the philosophical idea that (even the slightest, statistically negligible, and politically almost irrelevant) possibility of nuclear annihilation is incompatible with our moral constitution, as is in a different chapter our passivity in front of the worst consequences of global warming.

4. THE ORIGINAL GLOBAL
CHALLENGE: NUCLEAR WEAPONS

What features exactly global warming and nuclear weapons have in common, and how far they represent two different paradigms of global challenge will be discussed all throughout the book, particularly in chapters four and seven.[20] This complex matter cannot be anticipated here.

To lead the reader into a substantive understanding of what a global challenge is, nuclear weapons are the topic of choice. They are, first, the original global challenge, born 1945 in both reality and the cultural awareness of it, while global warming was discovered thirty years later and is not even now recognized as a global concern. Then, we knew a lot about the politics and the history of nuclear weapons, but eighteen years after the end of the Cold War the public knowledge of what the nuclear threat is like can no longer be assumed to be widespread. On the other hand, almost all of the available knowledge comes from that time. This makes a brief recollection of essential facts and meaning useful at this point.

Now, I shall not deal with nuclear physics and technology in general, nor shall I discuss what, at the optimistic outset of the nuclear age, was called their "peaceful use"—or "atoms for peace." On a general level, it is quite possible that nuclear physics, as Dieter Henrich (1990) has suggested, as well as molecular biology, I would add, have relevance not merely for moral philosophy, but also metaphysics, because both research fields confront us with the elementary powers which dominate our universe and our-being-in-the-world. But I am here not in the business of updating the philosophy of nature, a suggestive, but precarious enterprise, as its nineteenth-century evolution from Hegel to Engels makes clear. Nor am I talking about "nuclear power" in general. Nuclear weapons are distinct from, and more significant than nuclear power plants, because they are related to fundamental features of politics, like state(s), war, peace, and security. They are focused upon here not because of their physical structure (and arguable metaphysical significance), but as the last technical embodiment of those political categories.

Of the "weapons of mass destruction" devised during the previous century, nuclear weapons have been the most successful, in both meanings: they have been developed up to a much higher level of destructivity than, say, chemical weapons, and they have escaped, despite all arms control treaties, the international prohibition that has banned biological weapons, and recently chemical weapons too. There is no similar agreement for nuclear weapons, nor any serious intention among its holders to start thinking about it. At the end of the Cold War there have been significant arms reduction agreements, but not a first step towards general disarmament. Ownership of "nukes" is regarded as fully legal as far as it is limited to the original members of the nuclear club, the five powers with veto right in the Security Council, and as half legal for new *de facto* members such as Israel, India, and Pakistan. The non-proliferation regime has failed to fulfill the promise to make efforts towards general disarmament, which embellished the Non-Proliferation Treaty of 1968. Were all these reasons not sufficient for separating nuclear weapons from the other WMDs,[21] the simple fact that a nuclear war would involve the entire planet, while chemical or bacteriological warfare would be used only on single targets, can suffice.

The "definitive" character of nuclear arms is of course provisional, and will last only as long as a politically more effective weapon is devised, which theoretically needs not be an arm of mass destruction: let us, for example, speculate on a very selective weapon, capable of hitting the political and military command structure of the enemy, not his population. But such a pure Clausewitzian arm, devised to bend the adversary's political will[22] through direct impact on the bearers of this will, is mere fiction as of now, and will perhaps remain so. Furthermore, even in the case of a significant advance in military beam technology, explosions generated by nuclear fission or by a combination of nuclear fission and fusion seem bound to remain the main *modus operandi* of the weapons to come;[23] explosions affecting the military and industrial power of a country or even its people will remain the most effective way, to put the adversary under pressure, at least as a last resort. The idea that a shield is better than a sword, first proclaimed by Edward Teller and Ronald Reagan at the outset of the U.S. efforts to build an anti-missile defense system (Teller 1987) in the early 1980s, has not made in twenty-odd years significant steps towards realization, as far as the defense of an entire country is concerned. A partial shield, which might be able to protect a country's political and military headquarters and is said to be feasible, would not change the situation and could instead make it worse, by inducing the enemy to put full and continuous pressure on the unshielded population as the only way to coerce the decision-makers: only a "countervalue" strategy would be viable, once "counterforce" has become ineffective due to the shield.[24]

Since the very beginning of the nuclear age, marked by the explosion of the first (unexperimental) A-bomb over Hiroshima on the 6th of August 1945, there can be little doubt about nuclear weapons representing a global concern, the prototype of all global concerns. But the studies made in the 1980s, as the last nuclear arms race (around the so-called Euromissiles) between the United States and the Soviet Union stimulated a new wave of research, have shed a new light on the globality of this issue, if we take globality in the meaning defined above. Beyond the effects of nuclear explosions (blast, heat, radiation), which were already well known, the "nuclear winter" theory as well as other studies which do not ascribe themselves to this theory (though they have been to an extent stimulated by it)[25] have revealed the huge climatic and environmental impact of nuclear war. In its core, this theory maintains that a large-scale nuclear war (5,000 megatons on the whole, or 100 megatons concentrated on urban areas) would cause, as a secondary effect among others, a long darkening of the sun caused by smoke, dust, soot, and tar. The subsequent failure of the greenhouse effect in the lower atmosphere, which alone makes life possible, would let the average surface temperature fall by 20 degrees centigrade, according to the revised version of the "nuclear winter" theory (Sagan and Turco 1990, 193–96, 322). This phenomenon, along with the failure of photosynthesis through a lower light level and the radioactive and toxic soil contamination, would make the recovery of agriculture extremely difficult for a significant amount of time, thus heavily impinging on the life chances of both the survivors in the targeted areas and the inhabitants of non-combatant countries—even more so, if the nuclear exchange occurs in spring or summer, a circumstance which would immediately cause the loss of the harvest.

The studies which are not thematically related to the nuclear winter theory do not convey a less dark picture. As to the consequences for human beings, those not killed by the direct impact of the explosions would be confronted with enhanced percentage of cancer (World Health Organization 1984) and the weakening of the immune system. Their children, as far as those irradiated during the third and fourth month of gestation, would suffer teratogenic effects as well as mental retardation. Genetic consequences cannot so far be predicted, except for recessive mutations due to the in-group reproduction ("in-breeding") of small groups of survivors.[26]

Those who may survive in the targeted areas of the globe, as well as the inhabitants of "neutral" countries, would then suffer from three major sources of evil: the collapse of medical care (i) would enormously raise the proneness to diseases, epidemics, and mortality; the climatic and environmental changes (of both the natural and social environment) would impose (ii) a severe adaptation stress;[27] last, but by no means least, the collapse of agriculture as well as the industrial production and distribution, not to speak of the communication and finance systems, would induce (iii) malnutrition and starvation-to-death crises on a worldwide scale.

How many would have died in a Cold War turned hot?

In a full-range nuclear war (10,000 megatons, 90% of which detonated over Europe, Asia, and North America), the World Health Organization (1984, 22) reckoned with 1,150,000,000 dead and 1,100,000,000 people injured; this is more than one third of the present world population, which was in 1984 notably smaller. But the subsequent collapse of the medical and humanitarian aid, the longtime lack of food supply, due to the disruption of agriculture and trade, as well as other factors like the "struggle for life" between survivors, would have much worse consequences: the estimate varies from 1,000 to 4,000 million victims, independently of how many may have already been killed by direct impact (Dotto 1986, 116 and 128). On the other hand it has been calculated that the "human carrying capacity" of the earth, that is, the amount of food produced by natural systems without agricultural transformation, and presumably after the disruption of agriculture, would be able to feed just less than 1% of the present world population (Dotto 1986, 112).

In this book, the basic meaning of globality is that everybody on the planet is affected. Even in the unlikely assumption[28] that no nuclear explosion would occur in it, the Southern Hemisphere would be certainly hit by nuclear winter effects originating in the Northern Hemisphere such as decreasing temperature and rainfall. This would be aggravated by radioactive fallout and increased ultraviolet radiation, as well as the backlash on the South of economic and social disruption in the North (Sagan and Turco 1990, 171–76). The concern for nuclear war and disarmament is sometimes seen as a selfish interest or intellectual luxury of the affluent countries, whereas the Third World is more realistically worried about famine and oppression: a rather gruesome myth, circulated by leftist populism and the ideology of *tercermundismo*. The collapse of the North in a nuclear war can but have an appalling impact on the economy and the civil society of the South. But in the light of the nuclear winter theory that myth lacks even geographic basis. "Global" means here first of all "planetary." "We hold these truths to be self-evident. All men may be cremated equal," folksinger Pete Seeger's parody[29] of the famous passage from the American Declaration of Independence, must be taken literally, though the majority of them would not be cremated immediately by nuclear explosions, but rather would die later by disease and famine. Before this possibility was created by the nuclear build-up in the fifties, humankind existed as a philosophical assumption, not as a real element of agency. It now exists, if not as one actor, as one object of action (Anders 1956), be it a nuclear exchange or a catastrophic turn in climate change.

5. EXTINCTION?

Does the globality of the nuclear threat go beyond quantitative totality, that is nearly all men and women, or better, all presently living men and women? Does it involve the very survival of our kind, the chance for future men and women to come into existence? In the scientific literature the answer is uncertain, and cannot be but uncertain, as was the case with Albert Einstein, who first ruled out and later admitted the possibility of extinction (Sagan and Turco 1990, 70–71). The obvious impossibility of experiments as well as the complicated synergisms that could occur among various effects of large-scale nuclear explosions make this matter unpredictable, insofar as an unequivocal answer is sought. For the purposes of this book, it is sufficient to know that, while nobody predicts our extinction as a certain or highly probable event, nobody denies its possibility, as there are too many reasons for making it a scientifically conceivable event. This eventuality is either not discussed, as it happens in the official publications (United Nations 1989, World Health Organization 1984, National Institute of Medicine 1986); or it is discussed as a result of the long-time consequence rather than of the nuclear exchange itself.[30] Nuclear war and the subsequent winter would certainly bring about a large extinction event among plant and animal species, and some see high chances for *Homo sapiens* to be included in this event (Rowan-Robinson 1985, 80); others believe that it cannot be excluded (Ehrlich 1984, 10) and that we, after weighing the pros and the cons of this hypothesis on the ground of our present ability to predict, must admit that we are not able to determine the outcome (Sagan and Turco 1990, 74). Others still think that extinction will not necessarily occur, since small groups may be able to survive (Dotto 1986, 128).

    I have dared to assemble some verifiable natural science literature because I do not love to found my philosophical and political arguments on an imprecise and wholesome view of things to come. Now, the first philosophically relevant output of that recollection is that humankind has created devices whose use in a large-scale war would wipe out at least its largest part, even if total extinction may be not very likely; in any case a vast, almost complete destruction of civilization, both in its technical and intellectual/moral aspect,[31] is certain. A last scientific facet of this topic has now to be briefly examined: the (ambiguous) use of the term extinction.

Extinction is the usual end of the plant and animal species living on earth. Recent research has stressed the importance of catastrophic events in causing mass extinctions, as if in the old Cuvier *versus* Lyell dispute we ought to take sides for the former (Raup 1984). Louis and Walther Alvarez have advanced the geologically argued hypothesis that one of those catastrophes (which at the end of Cretaceous erased 75% of the existing species) might have been caused by the impact of an asteroid[32] or comet, unleashing a chain of events not unlike the chain called nuclear winter. This cataclysm, which eliminated big reptiles like dinosaurs (or were they birds?) and opened the way for the mammals' evolution to the human species, would thus have some analogy with the true nuclear winter, which may put an end to our natural history. The relationship between the extinction of organisms and our possible extinction does not go beyond this pale resemblance. Even before it is proved morally untenable, as I will argue in chapters five and six, the argument that extinction is in any case our destiny and therefore we need not worry about our extinction by nuclear war or environmental disaster, or even that we have no right to do so, is flawed from a naturalist point of view for the following reasons.

First, the differences between *Homo sapiens* and other animals are too many and too deep as to assume a sole and homogeneous type of evolutionary process towards extinction. Whatever anthropological definition of the human[33] we may want to choose, we have modified our environment (an important cause of extinction of plants and animals) and our conditions of living to an extent so far unknown to other organisms.[34] Is the fact that we are the sole species that has become nearly capable of bringing about its own extinction not an argument against regarding our extinction as a mere application of the general pattern of extinction?

Secondly, the idea that our extinction is a "natural" fact, and even that we "owe" our extinction to nature and the well-being of other species, is devoid of any scientific fundament and rather a deep-ecologist prescription or value judgment. Not only is the conventional Darwinian wisdom about extinction improving the adaptive level of the survivors no longer uncontested (Raup 1984, 3). Even more questionable is the idea that the world would be better without the human species; it is true that this has filled it with poisoning devices and subproducts such as nuclear and chemical waste, but it is *Homo sapiens* that is alone capable of servicing them, thus reducing risks and damages to all species, provided he and she put in their actions more rationality and a different rationality than they have done so far.

In any case, the extinction issue should not be overstated and reified, as if only the total physical disappearance of our species matters. Not extinction, but rather the destruction of civilization,[35] along with the death of thousands of millions of men and women, is the probable outcome of a large-scale nuclear war, as is the unlikelihood of rebuilding it in a way that has some continuity with our civilization and the meaning we put in it. At this stage I am stressing the importance of civilization in its technical and social components, regardless of any moral significance; what is of interest to me is the basic anthropological notion that for millennia human beings have received what they need for their biological survival only by means of civilizational organization. Bears need woods for their life, humans civilization; disrupting civilized coexistence is like burning down the woods.[36]

6. GLOBAL CHALLENGES AND GLOBALIZATION

At first sight the two notions converge, as both signal processes that tend to affect all of the dwellers on earth and to make them one addressee or perhaps one actor. We are also tempted to refer global (adjective) challenges to globalization (substantive) as one of its parts. There is nothing stringent in all this, and in general a relationship between two items is best described after first clarifying their differences, which are in our case remarkable.

In recent *history* the world has been so to speak unified by global challenges (1945, the 1970s for global warming) well before than by the present wave of economic, technological, and cultural globalization. With regard to the political nature of the two processes, global challenges have the potential (no more than the potential) to constrain citizens and states to act together or even to unify out of fear and desire of survival, while globalization can at most induce them to seek common regulations providing global governance. As to their effects, global challenges, once they are perceived as such, lead to common actions (non-proliferation regimes, international agreements for the protection of the atmosphere such as the Montreal and Kyoto Protocols), while economic globalization has as a main effect a mix of homogenization and fragmentation, which often results in several "glocal" configurations of the way particular communities adapt to it. This includes attempts to delay, stop, or reverse globalization that may also end up in neotribal ideology and behavior; they are however not unknown where, say, nuclear weapons are perceived as a token of national pride and power rather than as a global threat (remember the Pakistani crowds celebrating the country's experimental explosions).

These differences in the effects have to do with the different patterns of causation: globalization unfolds by osmosis, affecting the world little by little and generating a variety of reactions in the various local cultures and societies. The way of acting of global challenges on the opposite is systemic and sets off about the same physical effects all over the planet at the same time, for example after a large nuclear war or once a threshold in global climate disruption is arrived at.

There is lastly an epistemological difference: a global challenge is a descriptive concept pointing at two physical facts bearing directly on the political order in which we live. Globalization is a theoretical concept aiming at interpreting in a unifying manner various economic, social, cultural, and political processes. This is one more reason for not merging the two items into a general overarching tendency towards the global. This exists in some way, and our era can be generically seen as a nuclear and global era, but behind this formula lies a sea of difference, which is more important than the similarities if we want to have a credible picture of the present world. We ought then to look at the interactions between the two "global" items, but this cannot be done in this book, whose ambitions must be prevented from overreaching. Also, the future of economic globalization is at this moment not only fairly unpredictable because of its several failures[37] in keeping its promises and the numerous blocking or counteracting forces that have emerged; it does also require plenty of empirical research rather than philosophical speculations. This is why the reader will not find here a general theory of the link between globalization and global challenges. Specific links that may come up while examining global challenges will be obviously spelt out, for example in chapter seven the idea that what is called globalization of human rights should culminate in the recognition of humankind's right to survive.

NOTES

1. I am aware that this notion, which I started using in my writings of the early nineties, is now widespread throughout the debate on globalization in a fairly generic meaning, being for example the heading of a weekly report on CNN International. As a challenge always contains a threat, I shall choose between the two words according to the stress I want to put on the prevalently passive (threat) or rather active and inventive (challenge) reaction of the affected people.

2. Although frequently mentioned in this book, see for example at the end of § 5 below, the link of biological and civilized survival will be completely clarified only in chapter five.

3. Gérard Chaliand (2006), an expert on armed conflicts, believes that its impact will not be greater than that of the anarchists who inspired so much alarm at the end of the nineteenth century.

4. In choosing this method I draw some inspiration from the "materialistic" attitude of the early Critical Theory, especially Horkheimer's, as well as from Adorno's suggestion to look at the "experience contained " in concepts (*Erfahrungsgehalt*).

5. More preview of the single chapters is to be found in the introductions to Parts One (p. 23) and Two (p. 131).

6. Not to speak of other global issues included in the Copenhagen Consensus, such as financial instability, malnutrition and hunger, migration, and more, cf. Lomborg 2007. Of the threats mentioned by Rees 2003, a work which would require a more systematic discussion, those resulting from nanotechnology, high energy or artificial intelligence experiments are either not yet well known or unlikely to come into being, while the problem of unrepelled near earth–objects is not fully global, cf. note 32 below.

7. As poverty is defined in relation to national indicators of wealth, being poor in Europe or the USA is in absolute terms incomparable with the same condition in Mali or India.

8. This argument was originally raised in 1974 by Garrett Hardin (Hardin and Baden 1977). I do not make reference to Hardin's text, rather to the standard idea of the impossibility (or even illicitness) of altruism because of its devastating sub-effects given the limited resources of our environment and—what is more relevant to our discussion—the availability of a discriminating solution. Discrimination is always possible in the "lifeboat" argument, and consists in pitting those on the lifeboat against those in the waves, or the present lifeboat's occupants against their own posterity, which could be unduly disadvantaged, if their forefathers accept more and more people coming aboard; but it is by definition excluded in the case of real global problems, in which the beneficiaries, or the victims, are either all or none.

9. Why to use this notion almost as a namesake for "global challenges" will be explained in chapter five.

10. See a brief reconstruction of those not so fantastic fantasies, which reached climax on the CIBA-Symposium of 1962, in Flämig 1985, 27–31.

11. Let us however not forget that, for example, a medical advancement like amniocentesis is already being misused for tradition-oriented eugenics in India, where the early reconnaissance of sex provides the basis for a massive selective abortion of female fetuses (Hirsch and Eberbach 1987, 246–49). Yet, both this phenomenon and the use of Nobel laureates' sperm for the creation of superkids do not cross the border which separates the instrumental use of genetic knowledge from the manipulation of human germ-line cells.

12. Clones are genetically exact copies of the same individual, which can be diversified only by environmental influence. Chimeras are organisms created by combining cells taken from different embryos, even from those not belonging to the same species.

13. Cf. R.A. Weinberg quoted in Davis' final remarks, in Davis 1991, 264.

14. At the end of 2005, Moscow newspapers, cited by Stephen and Hall 2006 in *The Scotsman*, told of a request in this sense made in the 1920s by the Politburo to the world-renowned biologist Ilya Ivanov.

15. It is on this basis that every SS officer was declared to be "honor bound" to take part in Operation *Lebensborn* (cf. Hirsch and Eberbach 1987, 234).

16. Cf. Habermas 2003, particularly his notion of "humankind identity" (*Gattungsidentität*). Cf. Scott and Baylis 2005.

17. Unlike with nuclear weapons, I am not giving in this chapter a detailed introduction to global warming, referring the reader to §§ 1–2 of chapter four.

18. That nuclear proliferation (and not nuclear weaponry in itself) is "the most long-range and insidious threat to global survival" is the view of no less than Henry Kissinger (2004).

19. Mutual Assured Destruction.

20. To my knowledge only Thomas Schelling (2002) has once mentioned jointly the two issues with respect to analogies in the way they can be politically handled.

21. This notion and its short have again become successful in the ideological preparation of the war waged on Iraq by the Bush administration in 2003, but its analytical sense remains questionable.

22. According to the definition given at the beginning of *Vom Kriege*, war is "an act of force to compel our enemy to do our will" (Clausewitz 1832, 75).

23. Nuclear directed-energy weapons (NDEW) or third-generation weapons, which have been designed or even experimented with in the previous decades, employ beams, lasers, or microwaves in various manner, but all of them finally result in releasing energy generated by fission or fusion.

24. The two main nuclear targeting options are directed against the armed forces and political leadership (counterforce) or the economic and social assets of a country, beginning with its population (countervalue).

25. The "nuclear winter" theory, first introduced by Crutzen and Briks in 1982, was then developed by the so-called TTAPS (from the initials of the authors: Turco, Toon, Ackerman, Pollack, and Sagan). Sagan and Turco have later updated the doctrine, see Sagan and Turco 1990. A history (with large bibliography) of the debate, focusing on its strategic implications, is in Rueter and Kalil 1991. Sympathetic with TTAPS is Rowan-Robinson 1985. I have also taken into account four institutional publications, which are not committed to the "nuclear winter" theory: United Nations 1989, World Health Organization 1984, Oftedal 1986, Thompson 1986, and Dotto 1986. The last one is an authorized "popular account of the larger study" on the environmental consequences of nuclear war, written by a Committee (SCOPE) of the International Council of Scientific Unions. To my knowledge the nuclear winter (or its successive reformulation as nuclear autumn) doctrine has never been proven substantially wrong and still defines how we can provide a picture of what nuclear war would be like, as indicated by recent studies by Owen Toon and others on the climate changes generated on a *world-wide scale* by *regional* conflicts between small or medium-sized countries, each using fifty of their Hiroshima-like warheads, cf. www.colorado.edu/news/releases/2006/427.html.

26. Cf. Oftedal 1986, 344; *contra* and more pessimistically World Health Organization 1984.

27. See Thompson 1986 and Lifton 1968.

28. Unlikely because of the nuclear proliferation in the Third World, which in the last years has made a nuclear war more likely there, e.g. between India and Pakistan, than between developed countries.

29. In the song *To Be or Not To Be*, words by Vern Partlow, music by Talking Blues.

30. As Ehrlich 1984, 209, points out.

31. A destruction of both *Zivilisation* and *Kultur,* as German philosophers once used to put it.

32. This danger seems to surface again with the asteroid called Apophis, which in 2036 could hit the earth with catastrophic effects (probability 1:45,000), cf.http://www.esa.int/gsp/ACT/mission_analysis/asteroid.htm.

This is only half a global challenge, as the countermeasures that could be put in effect do not necessarily need the cooperation of all countries. Should however the threat be confirmed by further inquiry, not to take them would raise problems of guilt by omission towards the next generations, a topic related to that discussed in chapter five.

33. Such as "tool-making animal" (Benjamin Franklin and Karl Marx) or technically oriented "deficient being" or *Mängelwesen* (Arnold Gehlen). On how the criteria separating humans from animals have become relative or fuzzy, see Fernandez-Armesto 2004.

34. Margaret Mead has pointed out that the attitude, which Gehlen calls the technical approach of humankind to the world, derives from its particularity towards other species, inasmuch as human beings need protection as much as other animals, but they have no built-in ways of protecting themselves against the effects of the enormous size of their interventions on nature (cf. Mead 1964).

35. Or, in the case of the disastrous consequences of global warming, civilization would be vastly upset by unimaginable consequences in our physical environment.

36. More on this in the introduction to chapter five.

37. Last came in the summer of 2006 with the collapse of the Doha round in worldwide trade negotiations.

Part I

THE RISE AND FALL OF MODERN POLITICAL RATIONALITY

In this part I argue that the only two challenges that can be regarded as truly global in political theory, that is nuclear weapons and global warming, affect present humanity and their future to an extent that cannot be addressed by applying the standards and notions adopted to govern the polity since the inception of modernity in the sixteenth and seventeenth centuries. They require politics and its relationship with morality to be redefined as I have set out in Part Two.

Two preliminary steps need to be taken in order to argue this main thesis. First, we need to eradicate what I think to be a misleading image of late modernity as an era defined by the predominance of risk, to which we have to adapt, whatever the risk, social uncertainty, or nuclear threat may entail. This theory contends that the life of men and women, both as individuals and collectively, has always been a journey with an uncertain destination, as the writers of ancient times already knew. Furthermore, all religious, metaphysical, or "scientific" (in a positivistic sense) attempts to give a definite and risk-free character to our destiny have proved unsustainable and have been successively dropped in the modern era. We should thus not dramatize the global challenges, a skeptical observer would suggest, and accept them as nothing more than the present shape of the eternal uncertainty of human destiny. In chapter one I will reject this view in the belief that, if we take the notion of risk seriously, we cannot apply it to global challenges, which require a political and moral investigation in their own right.

In chapter two I will explain what I mean by the notion of political modernity. Modernity in general, that is modernity in philosophy, science, literature, and the arts, is a sufficiently common and well-understood image. Not so *political* modernity, which is mostly associated with domestic features such as

23

constitutionalism, liberalism, and democracy, while the external aspects of the Leviathan, that is, sovereignty, diplomacy, and international anarchy, are left to International Relations literature. It is as if social justice and political equality, or the doctrines and movements upholding them, were more relevant to the definition of the modern era than the amount of death, suffering, and destruction inflicted by warfare upon millions of men and more recently women. Because of these different approaches no overarching and generally shared image of political modernity is available at this stage of research, nor, even less so, one of political modernity as a rational order. Rationality in modern politics is mostly understood as a claim raised by authors determining how they want to shape the society of the future, rather than as a quality of present or developing institutions such as the international balance of power. This confusing plurality of meanings of "political modernity" and "political rationality" is the reason why in chapter two I deem it necessary to go at some length through the various elements of these notions in the relevant literature and to define my own version of them; this is vital for the understanding of my main thesis on the end of modern political rationality. I do not stop at this kind of rationality, but include the rationality claim of modern science in my argument, while also discussing the bipolarity of instrumental vs. encompassing rationality, of *ratio* and Reason. However, the insistence on *political* rationality is legitimate with regard to my second thesis, which sees the crisis of modernity unleashed by global challenges as a result of a lack of political institutions that might be able to come to terms with them.

On the basis of these preliminary clarifications, chapters three and four deal with the two different global challenges as something which, each in its own way, is putting an end to modernity as a somehow rational order and urges us to rethink the very fundaments on which we regulate our life as a species on this planet. Chapter three looks at the emergence of the potentially suicidal nuclear weapons from the unlucky encounter of twentieth-century physics with total and ideological war, and the resulting demise of the international institutions of modern rationality, starting with Clausewitzian war as a feature of politics. It also indicates how the science and technology whose use can be so threatening stems from the same rationality that inspired the project of political modernity. Nearly twenty years after the Cold War made a first step towards its end with the first nuclear disarmament pact, the INF Treaty of 1987, the essential features of nuclear bipolarism are unknown to younger readers and need to be remembered in some detail, as I shall do in chapter three; also, we still lack a philosophical reconsideration of what that period meant to humanity and the attempt made here needs historical foundations.

The same considerations about science and technology, which include the end of their moral neutrality, hold true for the man-made climate change.

Chapter four shows how, ontological and epistemological differences between the two items notwithstanding, global warming and our political impotence in the face of it confronts us with the same lack of timely intellectual and institutional change which can be seen in our stand towards the nuclear challenge.

Chapter One

Risk or Challenge?

In conversations or newspaper articles we are used to expressions of common parlance such as "the risk of nuclear war" or "the risk of devastation of coastal areas caused by global warming." They are implicitly regarded as the most severe among the various risks that overshadow our life on this planet. The idea that nuclear war and disastrous climate change are homogeneous to other, even if far less dramatic risks is implicit in the conventional wisdom according to which we now live in a "risk society" and have to come to terms with this fact.[1] This view is different from this book's approach to nuclear weapons and man-made climate change as specific, ultimate, and incomparable threats, and needs to be discussed preliminarily in two steps, corresponding to the two sections of this chapter.

1. ARE GLOBAL CHALLENGES RISKS?

There is admittedly some truth in playing down the otherwise usual dramatization of new risks. Had Columbus, his crew, and his Spanish backers not decided, consciously (though confusedly) and purposely, to take risks, the undertaking of which, for better or for worse, ushered in the modern era would have not taken place. Another mariner, Robinson Crusoe, who had survived many risky situations, was even persuaded that life can be managed better if we do not know or think too much about risks: hidden risks should be preferred to the overt ones (Defoe quoted in Douglas and Wildavsky 1982, 28). Modern society as a whole is indeed widely based on *risk-taking* and *risk-managing*, because the growth of social complexity (as a consequence of the rise of the world market) and a technological exploitation of nature never seen before have made them a specific feature of modern life. A similar picture can be extracted from Max Weber, as he stresses the prevailing role, in modern economy and society, of instrumentally rational action[2] which includes a calculation of risks and consequences, against the traditionalistic (reaffirmation of old values) and the charismatic (visionary innovation) patterns of action. In both the instrumentally rational and the charismatic patterns, we act while looking towards the future, but only the former is able to make risk-taking a known and manageable element of action. Acting on value rationality *(wertrationales Handeln)*, the other approach to modern rationality, also takes risk into consideration (otherwise it could hardly be called rational), but only to state the superiority of the pursuit of values in risk calculation. Besides its domestic and social aspects, as analyzed by Beck 1991 and Leiss and Chociolko 1994, this basic pattern can also be seen as the keystone of the international "balance of power" system, whose dynamics of changing alliances was devised in order to take into account and prevent the risk of preponderancy: risk management is an important example of the self-sustaining equilibrium of the modern system of states.

Global challenges confront us, however, with risks that outgrow the modern understanding of risk. But let us first define what we mean when talking about risk. Out of a large array of definitions, we can define *risk* as a function, that is, the product of the probability and size of loss (C. Vlek and J.P. Stallen quoted in Brehmer 1989, 27).[3] But risk assessment is not a technical matter, a relation among abstract entities. At least in our context, risk twice involves human agency (cf. also Luhmann 1988, 96–100). We only risk as far as we act; otherwise we should not speak of risk, but more generally of menace; in other words risk is always risk-taking, even in the shape of doing nothing in certain circumstances. Secondly, the selection of the damage we are ready to accept by acting in one way rather than in another is culturally determined, as it is always made in comparison with the net benefits which we expect from the risk-taking activity under scrutiny (Leiss and Chociolko 1994, 11).[4] This is not to agree with cultural relativists on risk being nothing else than a social construct, because social constructs can and in this case do leave physical deposits (fissile material, excess greenhouse gases) which can be measured just as well as they can kill or hurt generations of humans.[5] In other words "the estimation of a risk is a scientific question, its acceptability a political one" (Handler quoted in Douglas and Wildavsky 1982, 65).[6]

Now, what problems do we encounter when facing risks? We have to strive with two kinds of *uncertainty*: one regards the probability of a loss, the other our evaluation of its size or significance.[7] This conceptual scheme looks like it could be applied to global challenges too, thus leading to the conclusion that, since the probability of a catastrophe is so tiny as to compensate the huge size of the loss, we need not put environmental and nuclear risks at the top of our list of preoccupations. However, the substantive peculiarity of global challenges means risk thinking is not applicable to their case. Let us see why.

The probability of the worst outcome (extinction of the human kind or annihilation of a great part of it by nuclear war or catastrophic environmental changes) can by no means be assessed in such a way as to assign numerical values to the various degrees of probability, in correspondence to various courses of action. Tendencies in global warming can be to an extent predicted by extrapolating present figures of energy consumption and related phenomena, but these predictions are still far from certain, as we shall see further in chapter four. Insofar as they do not involve merely physical developments, but also the political actions that may unleash or moderate them, all numeric approaches to prediction and probability lose credibility, because of the unquantifiable nature of complex political processes. The risk or rather the uncertainty of deadly outcomes exists and cannot be said to tend towards zero, even if we may indicate it by a very small positive number.

Let us now look at the second element of the "risk" function, that is, assessing the size of the loss. As we do not mean merely the number of dead, but the *moral value of the loss* as well, its extent depends totally on the criteria we use. From a philosophical vantage point, it is neither self-evident, nor beyond all discussion that the loss of a large part of humankind or its civilization would be an unbearable, incommensurable loss, which must be in any case averted. This position must be argued against divergent options, such as nihilism, as I shall try to do later at the end of chapter five. But, though "absolute" values may be philosophically questionable, this is still different from the problem of the method we want to employ in order to assert a value. My preoccupation is that, if wide-ranging political consequences are to be drawn from the global challenges, they should not rely on values which pretend to be intuitive, or decisions based on the preferences expressed by the actors, as is usually assumed in decision theory. Our own or our posterity's preferences are unstable because they are subjected to all possible shifts in our constellations of values: not only in the turnover of the generations, but already in one person's life span. We can for example decide to build a new highway in order to speed up traffic and/or to lessen traffic-related casualties on the existing road; or we can decide to abstain from doing so, thus saving the wilderness. It is a cultural choice between values (better transportation within the given order vs. a different attitude to nature and our self), which in either case results in divergent assessments of the risks entailed in the two courses of action. The choice we are making today is possibly different from that which we may have made twenty years ago, and our great-grandchildren may change it in the framework of cognitive and practical preconditions whose knowledge we cannot anticipate.

Now, this *relativity* of choices about values can hardly be the firm ground on which an undertaking as vital as the survival of humankind and civilization can be based. Should we then re-invent an objectivistic theory of value, risk, and decision in order to find that ground? Quite on the contrary, I think that we can locate it without leaving the path of the self-reflection of culture. Our decision to, say, take adequate steps to prevent the unleashing of a nuclear war is a cultural choice too, but the related risk assessment seems to be less dependent on variable value orientations. Only by eliminating the risk of being blown out as humankind will we preserve the setting (a risky world) and the agents (risk-prone beings) in which and for whom risk-taking makes sense. Contrary to the advice to "take it easy" with the global threats, primarily those resulting from nuclear weapons, it is the very proneness of modernity to risky endeavors that requires us to get rid of a risk which seems preposterous to keep calling a risk, since it would eliminate all preconditions of risking and experimenting. As with other aspects of modernity, or rather civilization in general, to recognize risk-taking as one of its important features does not dispense us from recognizing that its dynamics may become self-defeating if we are not able to locate or respect the limits of what can be handled as a risk.

Since in this case its procedures are superseded by the size of the loss, decision theory does not provide much help in rationalizing our confrontation with global issues. Before we turn to further aspects of this impasse, let us nevertheless draw a conclusion within that theory: given the impossibility of calculating our risks in numerical terms, we have to recognize that we are not operating under conditions of *uncertainty*[8] rather than under those of risk. The decision theory tells us to follow the maximin criterion: if only sub-optimal outcomes are possible, then choose the best worst outcome. Between multipolar nuclear deterrence (worst outcome: nuclear war and winter) and, for example, supranational institution-building (worst outcome, in the less favorable version: nuclear peace under a worldwide tyranny), we have good reason to choose the latter. This example, which at this point is only a mental experiment, not an argument in favor of world government, is valid under the conditions that

a. no better option (like total and permanent nuclear disarmament) is available; and
b. the second option (the institution, not necessarily the tyranny) can really be made available, a presupposition which shall be discussed in chapter seven.

In the two further aspects of the impasse, substantive and epistemological issues are intertwined. To begin with, among options with uncertain outcomes it is usually considered a wise method to postpone making a decision, and first search for more evidence. But *time* matters in global issues, even more than it does in (local) environmental ones: this seems to be true not only with regard to global warming, but nuclear proliferation as well. It is not simply that over time the process put under scrutiny because of its level of danger keeps developing with cumulative effects, while we drag out our decision; a further reason for not suspending our decision-making is the time lag which in many issues may exist between making a decision and having its benefi- cial (maximinimizing) consequences become effective, thus preventing the undesired outcome of the risk from taking place: you make a right decision only insofar as you make it with the right timing, not merely giving to your decision the right content.[9]

The second feature of the impasse is *irreversibility*, which is peculiar to the worst outcomes of global challenges and to some more ordinary issues of en- vironmental policy as well, for example, the extinction of a species. We can- not completely undo the hole in the ozone layer (it will take decades to re- cover, even if we totally and immediately stop using chlorofluorocarbons); nor can we be confident that, after a large nuclear war, we would be able to reconstruct world society as we did after World War II. Not addressing the global challenges is not a risk that can be taken in the expectation that, if something goes wrong, we pay the price owed and go back to business as usual, or nearly as usual, as happened after Hiroshima and Chernobyl. The difference is—and this is the third aspect of the impasse—even greater, at least with regard to nuclear weapons: if something goes wrong, it could be not just "something," but everything and everyone that is doomed. Among the ca- sualties there would probably be the very actor (humanity as a civilized species) who calculated and decided to take the risk (even if the calculation and decision were actually made by few leading members of our kind, a fact whose relevance we will soon assess). This is a circumstance that is not con- sidered in any theory or philosophy of risk and is rather likely to outmaneu- ver this altogether. Whoever would counter this argument with reference to an established game like Russian roulette, should bear in mind that in this game

1. the player has something to gain, if s/he wins and does not lose her/his life (money, self-esteem, or social esteem because of her/his "courage");
2. if s/he kills himself, s/he only kills her/himself and not others (a collective version of the game has not been proposed);
3. others (family, group) could even reap benefit from the money or the fame s/he may leave behind.

None of these circumstances or opportunities apply to our risky game with lethal weapons. If we want to preserve our modern ability to rationally take risks,[10] we should not deal with global and ultimate menaces as if they were risks to be taken. There is nothing to gained by taking them. The unprecedented severity of the possible losses and the uncertainty in which these issues are enveloped request a different approach, which will be looked into in the last three chapters.

2. HUMANKIND AND RISK

There are furthermore other reasons that make risk theory non-viable in this case. We have so far introduced humankind as the actor for whom it makes or does not make sense to take the risks linked to global challenges. But until now humankind has been a fictitious actor, whereas decision theory assumes real, unitary, and rational actors. As I hinted above, decisions to possess nuclear weapons or to reject targeted emissions curbing are made by some countries and not by others, by one wing in the elites of those countries and not by another. This makes it impossible to regard all this as a sequel to the risk-taking, risk-assessment, and risk-management actions undertaken by an actor. Also, possessing nuclear weapons with all the relative benefits and risks is the effect of political decision, while for two hundred years generating man-made climate change has been an unwitting side effect. This is no longer the case since global warming was discovered in the 1970s, and non-acting on it now means assuming the risk of severe damage to one's own country, the planet, and future generations. *Humankind* has thus recently started its effective, if inchoate, existence as an actor, for example when the majority of the states agreed to curb the use of chlorofluorocarbons (Montreal Protocol) or to protect the seabed as "common heritage of mankind" (cf. chapter seven, § 4). On other issues such as greenhouse emissions and nuclear armament, however, so far humankind remains a single victim, as everybody can be hurt, but not a single actor, as it is rather split into many conflicting ones.

Let us lastly go back to the question of the expected *benefits*. The possibility of lethal outcomes is to be weighed against the degree of security provided by national possession of nuclear weapons, be it for war or deterrence purposes, and the degree of well-being provided by unrestrained energy consumption. The uncertainty of outcomes is high in both cases, because deterrence does not exclude war, and the knowledge of climate evolution patterns is too incomplete as to rule out the possibility of a dramatic shift in ocean circulation (cf. chapter four, § 2). Even more uncertain is the identity of the actor who should take the risk of such lethal outcomes: can a single nation act in its own (presumed) national interest, whereas the costs may be burdened on the whole of humankind? Can the present generation make decisions based on its particular understanding of security and well-being while their effects will primarily benefit or ruin future generations?

In this chapter I have argued in several ways that the idea that we can regard nuclear weapons and huge greenhouse emissions as risks among the many others in the "risk society" is fundamentally flawed, because for epistemological reasons risk theory cannot be applied to global challenges. That idea may make sense from the point of view of the sociologist, whose focus is on the widespread, if generic feeling of precariousness heeded by the people at the end of modernity and related to such varied issues as the job market and terrorism or nuclear threats. On the contrary, the philosophical approach pursued in this book does not take the present risk perception among the public[11] for granted, but shifts the attention to the meaning of the effective, verifiable threat situation for the polity and asks about its normative consequences.

This initial clarification was the preliminary step that needed to be taken in order to clear the way for a different and wider approach, which tackles security, a notion we have already met in this chapter, and tries to develop it in the full range of its meanings in history and politics.

NOTES

1. This notion can be traced back to Beck 1992, the book that successfully introduced risk theory into general social theory, but I am referring here to a standard view of it rather than to Beck's original version. More at the end of this chapter.

2. Or goal-oriented action (*zweckrationales Handeln*).

3. A good introduction to risk theory and its terminology can be found in Burgman 2005, chapter 1.

4. The question of the expected benefits will be discussed later in this chapter.

5. For a critical discussion of cultural relativism in risk theory see Shrader-Frechette 1991, chapter 3.

6. Political estimation of the acceptability of a risk is influenced by the different risk perceptions among elites, advocacy groups, and the general public, a sociological matter that cannot be analyzed here. Cf. Jasanoff 1993.

7. Hammitt 1990 speaks of "outcome uncertainty" and "value uncertainty." Elster 1979, 372–73, upholds the difference between decisions made under risk or under uncertainty: in both cases the outcomes are known, but only in the first can we also assign numeric probabilities to each of them. While discussing risk and decision theories, both authors, as well as Douglas and Wildavsky, use environmental issues as the main reference.

8. I borrow this development from Elster 1979, cf. previous note. I also share, although for different reasons, Elster's skepticism against the argument that he calls

"Pascal's wager turned on its head" (Elster 1979, 390): the risk attached to nuclear power (I would rather refer to nuclear weapons alone) is in any case infinite and should therefore be banned, because the product of even a very small positive number (the probability of nuclear war) by minus infinity (the value we are expected to assign to the extinction of our race and civilization) is an infinite number. Rather than Elster's fear of inflation (there are too many actions to which we could associate disastrous consequences), I would advance the conviction that ultimate philosophical questions such as God's existence or the survival of humanity cannot be addressed in mathematical terms. *Ne sutor ultra crepidam*, the Roman saying telling the shoemaker not to exceed his job, should be heeded by decision theorists as well.

9. Examples of right measures which were taken too late are provided by Parson 1993. As explained in chapter four, even measures mitigating greenhouse emissions and global warming may come too late, if it turns out that the critical threshold already lies behind us.

10. On how to approach the link between risk and rationality I share the view of Shrader-Frechette 1991.

11. To my knowledge there is little if any empirical research on risk perception regarding nuclear war, particularly after the end of the Cold War, and global warming. The latter is presumably unspecifically included in the inquiries on the perception of environmental risks, while interest in the perception of the nuclear war threat has all but disappeared since the end of the Cold War; still valid are contributions from the *histoire des mentalités*, cf. Weart 1988, a groundbreaking book, and psychoanalysis, cf. Levine 1988. While surveys on social values and issues such as family life, employment situation, and welfare abounded and still abound (cf. the WorldValuesSurvey, particularly the 1990 Questionnaire, carried out in a year still in the shadow of the Cold War), the lack of attention on the perception of threats of war and environmental disaster reveals the denial mechanisms that affect the choice of research topics no less than the awareness of the public in everyday life.

Chapter Two

The Rationality of
Modern Political Order

Risk-taking, risk-managing, and risk-avoiding are facets of a more comprehensive category: *securitas*, or security/safety.[1] The search for and the buildup of security is a central accomplishment of the *modern* social and political system, although it had already played an important role in Rome. It is also a keyword in modern political thought, at least among those theorists to whom the existence of political society is a problem rather than a given, as it is to both the Aristotelian and the Hegelian tradition. How and why the polity replaces the state of nature is a problem first of all to the contractarian current, from Hobbes to Rousseau, but they were not alone in raising the question of what makes a stable communal life possible. Security is indeed crucial to answering the question of what the societal, and eminently the political organization among human individuals is supposed to provide for their advantage, if they are to enter it and to remain in it, instead of staying aside or disrupting it.

The scope of this category (§ 1) is indeed not confined to the political realm, and involves other fields such as philosophical anthropology. In political and social philosophy, security means stabilization of the individual's life chances in a collective situation and through social interaction. In premodern stages of Western culture this kind of stabilization was not always sought in society, but also in the withdrawal from public activities and the cultivation of the individual mind, as in Epicurus' *ataraxia*; or later in Augustine's *civitas Dei*, which alone can provide true peace. The cultural and political unrest that characterized the twentieth century is mirrored in the increasing importance which the uncertainty and rootlessness of human life has achieved in philosophical reflection, like in Heidegger's word (1978, 174–75) on the "thrownness" (*Geworfenheit*) of the human being and in the existentialism of

35

the early Sartre. In German philosophical anthropology the precarious un-steadiness of our existence has been traced by Arnold Gehlen back to the def-inition of man as a "deficient being" (*Mängelwesen*), whose lack of special-ized organs in his or her relation to the world explains the need for technique to mediate this relation, as well as the role of institutions (in a broad socio-logical sense) in stabilizing behavior and expectations.[2]

How crucial the political regulation of security has been in modern times is argued in § 2 in terms of an unavoidable Hobbesian moment, which goes far beyond the conventional perception of Hobbes as a philosopher of state absolutism. Then in §§ 3–5 the evolution of Hobbesian categories such as fear, anarchy, sovereignty, common power, war, and the law of peoples is traced in the history of both legal and informal European institutions, rather than through philosophical literature. This amounts to a comprehensive pic-ture of modern rationality, whose bipolarity of instrumental *ratio* and Reason is illustrated in § 6. This chapter as a whole depicts the creation of the polit-ical rationality whose dismantlement is described in chapter three and also provides some historical background for the discussion unfolded in chapter seven about what may come after modernity.

1. POLITICAL SECURITY

This brief recollection of philosophical episodes is intended to visualize some points, rather than to introduce a general theory of security, which would be of little use:

a. Security is not a mere political category, and political security is connected to security as it is defined in other fields (among these, anthropology and psychology);
b. In its basic shape, security is a problem for politics at any time. In a (at least at first glance) less complex world than our own, external security concerns could be addressed by building *limes*, as the Romans did against German tribes, and internal ones by, say, killing potential rebels in a treacherous pre-emptive strike, as Cesare Borgia did after inviting them to dinner, ac-cording to Machiavelli's account,[3] or, like Shakespeare's Richard III, by sending Buckingham to the scaffold.

On the back of these considerations I however argue the *thesis* that secu-rity problems have a peculiar relevance in modernity, and are used to find an adequate answer in the order provided by the state, while it is either modern politics and state, as in the case of nuclear weapons, or a lack of political regulation of the economy and technology, as in the case of global warming, that makes the modern system of security insufficient and even counterproductive. It is therefore in the realm of politics that present and future insecurity must be addressed in the first place, whatever remedy philosophy or psychology or religion may provide it with. In turn, the political solution must also be able to address problems of insecurity which lie in layers of consciousness beyond the political realm.[4]

Both macro- and micro-historical evidence can be provided in relation to the idea that the need for security has attained unprecedented importance in the modern world. In less than two centuries a (Western) world order that had lasted over one thousand years was turned upside down: earth and man were no longer at its center, nor the Pope and the Roman Emperor at its spiritual and political apex, nor was the Mediterranean still the main stage for trade and civilization, as it was increasingly overshadowed by American and East Indies trade. After Charles VIII of France invaded the Italian states (1494) and Luther launched the Reformation movement (1517), wars of supremacy and religion ravaged Europe; on its Eastern border, after the fall of Constantinople (1453), Europe had to make at least three major efforts to stop the Turks (at Malta 1565 and Lepanto 1571 and then again at Vienna 1683), before Eugene of Savoy, *der edle Ritter*[5] in Hapsburg's service, was able to establish a safe Balkan border with the Ottoman Empire in the eighteenth century. In the intellectual sphere a further, sudden, and impressive widening of the old world came with the introduction of printing. In this framework, the idea itself of (Christian, European) unity, which in the Middle Ages had been supposed to provide security and justice on the basis of a universal creed, institutionalized in the Roman Church, and under a central, supranational power, the Holy Roman Empire, was definitively abandoned. In its place, separate political units, some of which already existed, consolidated their power in France, Spain, Portugal, Great Britain, Switzerland, the Netherlands, and later in Poland and Russia; and more importantly, the new pattern of order found its legitimacy in the Treaties of Westphalia of 1648.[6]

On the other side of this story, that is, looking at how the people experienced constraints put in place by system mechanisms or historical change, the subjective face of insecurity and threat, that is, fear, was a widespread feeling among millions of men and women before it surfaced as a concept in Hobbes' state of nature. At the dawn of the modern age in the Western world there was a soaring wave of fear across all countries and social groups—this is how the leading historian of fear and insecurity sums up his recollection of episodes and behavior patterns which cover most fields of everyday life and institutional regulations (Delumeau 1979, 20). More laden with political consequences than any other single phenomenon was perhaps the fact that, after the fall of chivalric ethics, fear became a *publicly legitimate feeling*. If it is true that the "process of civilization" substantially consists of the regulation of violence and fear (cf. Elias 1936), this regulation could only then become the cornerstone of a political project like the modern state as fear became a reflective aspect of a society which was not simply fearful, but recognized fear as one of its elements. Hobbes perhaps had this in mind, as he twinned his birth in 1588 with the Grand Fear that shook England in the same year as the Spanish *Invencible Armada* approached.

2. THE HOBBESIAN MOMENT

In Hobbes' materialistic view, reason itself is but the main instrument of self-preservation, and this is the outcome of a *ratiocinandi actum* of calculation about the consequences of our own actions. In the search for "convenient articles of peace," which may be instrumental to the care for our own life and body, the "natural reason" is driven by those passions which "incline men to peace," first of all the "fear of death" (Hobbes 1651, part I, chapter 13, 188). This is both the feeling of being scared (*perterreri*) as well as a permanent state of fear (*metuere*), which includes "diffidere, suspicari, cavere, ne metuant providere (to mistrust, to be suspicious, to beware, to act lest they have to fear)" (Hobbes 1642, sectio III, caput I, 161n.). More than actual fighting, it is the *metus* deriving from its possibility, due to the lack of a central power, that matters. Also, there is nothing in nature's spontaneous processes that tenders us any protection. This can only come from the correct use of the natural reason beyond the original state of nature, that is, from the build-up of a voluntary construct, of an "artificial man," "that great Leviathan [...] called a Commonwealth, in Latine Civitas" (Hobbes 1651, part II, chapter 17, 227).

For our present purpose it is not necessary to mention all the steps and problems in the genesis of the *civitas* or Commonwealth; this, by the way, and not "the Sovereign," is Hobbes' name for the state, sovereignty being the "artificial soul" of the artificial man, that is, the most important of its *modi*, not its substance. Hobbes' commonwealth provides domestic security, concentrating in its hand all power and force, as well as defense against external threats. Here lies the link between protection and legitimacy, which highlights protection as the fundament and limit of political obligation:

"The Obligation of Subjects to the Sovereign is understood to last as long, and no longer, than the power lasteth, by which he is able to protect them [. . .] The end of Obedience is Protection" (Hobbes 1651, part II, chapter 21, 272).

Another relevant point concerns fear: in Hobbes' state, it is bound to disappear in its original shape as the combination of *metuere* and *perterreri*, as we have seen. While it disappears as a generalized, but emotive reaction to the un-predictable dangers posed by the *bellum omnium*, it is now built into the rational-artificial organism of the state; in this no other fear exists except the fear of the law (*metus legis*), which in the contractarian tradition, not only in Hobbes' "absolutist" view, is regarded as compatible with freedom. But, even after the domestic fear has been transformed in this way, should not the citizens or subjects retain the original (state-of-nature type) fear, as far as external threats are concerned? Although not explicitly, Hobbes seems to deny it: the actors in interstate conflicts are states, not citizens; the fear is theirs and must be converted into rational policies of war-preparedness (Hobbes 1651, part I, chapter 13, 187). What happens if, nevertheless, the worst case occurs, and the country is invaded and looted by the enemy, as was usual in continental Europe during the Thirty Years' War, which ended three years before *The Leviathan* was published? Hobbes' disenchanted *ante litteram* utilitarianism leads him to furnish theoretical justification for what was a widespread tactic in his time: the citizens are not merely allowed, but rather obliged to deny their loyalty to the sovereign who has failed to accomplish his or her duty to protect them, and transfer it to the victor (cf. Hobbes 1651, part II, chapter 21, 272–74).

   In its first philosophical formulation—Machiavelli was no philosopher, nor Bodin so clearly modern—the core of modern politics is thus located by Hobbes in the build-up of an institution which reduces violence by concentrating it in its own hands, establishes peace and safety (*salus populi*), and whose legitimacy relies primarily on the accomplishment of this task. This basic act in the sense of security and order is not all that politics is concerned with, as it can be led by various finalities and projects, but it remains the condition politics must itself generate in order to make all its other dimensions possible. This is the "Hobbesian moment" in modern political philosophy, or in other words the nucleus of the covenant, which can also be found in Locke or Rousseau, and it still constitutes the first problem with which politics is faced, whatever our attitude to Hobbes' solution may be. This is why Rousseau requires "the total alienation of each associate, with all his rights, to the whole community" (Rousseau 1762, part 1, chapter 6, 138). But it is true as well of Locke's civil society, in which, on behalf of self-preservation, "the unalterable, sacred law for which individuals enter society" (Locke 1690, chapter 13, 193), "everyone of the members hath quitted this natural power [to punish the offender of the law of nature, explanation mine], [and] resigned it up into the hands of the community" (chapter 7, 159). As for the legitimacy created in this way, this view was even shared by the skeptical David Hume, who conceded that the sole foundation of the duty of allegiance to government is the advantage of preserving peace and order among mankind; a view that he extended to the intercourse among different societies, as regulated by the law of nations (Hume 1777, chapter 4). Whatever the picture of the (lethal or benevolent) state of nature, whatever the idea of (absolutist or limited) government, it is not with regard to the core answer to the question "how is it possible that human beings live in society?" that Hobbesian and Lockean views can be differentiated, as Bull (1977) and Nardin (1983) seem to believe.

Thomas Hobbes' view later developed into Max Weber's definition of the state as the political (i.e. dealing with the distribution of power) institution which is characterized by the successful claim of monopoly to the *legitimate* use of force in the enforcement of its orders within a certain territory (Weber 1922, 54); all later definitions are variations on Weber's.[7] By pointing exclusively at the *modus operandi* of the state, and dropping all substantive indications of ends, this definition of the modern state also reveals its own epistemological modernity. However, if we look back to what we assume to be its root, that is, the contractarian idea of human individuals renouncing their right to "diffident" self-defense (Hobbes) or to be the judge in their own cases (Locke), this process still seems to include a finality: to provide peace, safety, and the public good of the people, as Locke put it in chapter 9 of his *Second Treatise of Government* under the title *Of the Ends of Political Society and Government*. But, while giving more reasons (that is, some advantage for everyone) for people to live together rather than to split into warring factions or self-sustaining units without a state, peace and safety, in a word, *order*, are not a substantive finality, motivated by some particular ideology or group interest; they merely represent the basic conditions that make the pursuit of substantive ends possible. None other than Karl Marx and Friedrich Engels recognized that all non-primitive societies, in which division of labor prevails, have a need for order, that is, a need to recompose their segments of interaction into a framework of "cooperation," a true general interest which the ruling class bends to its special advantage, shaping the state as an instrument of oppression (cf. Marx and Engels 1845, chapter 1).[8]

A corollary to the "Hobbesian moment" is Hobbes' rejection of what we shall discuss in chapter seven under the headline of the "domestic analogy." States and their heads, always in "the state and posture of Gladiators," may be even more bellicose than individuals. But to individuals the threat is total: even "the weakest has strength enough to kill the strongest, either by secret machination, or by confederacy with others" (Hobbes 1651, part I, chapter 13, 183). On the contrary, even in the posture of war, states "uphold the industry of their subjects; there does not follow from it, that misery, which accompanies the liberty of particular men" (ibidem, 188).[9] The continuation of economic activity as well as the existence of civil society, if they can both be seen as included in what Hobbes meant by "industry," are not threatened by *interstate war*, certainly not to the extent they are by civil war: for the preservation of individuals a covenant between them is necessary, a societal tie between states is not. I would suggest that this element makes any commonwealth fulfilling the requirements of the "Hobbesian moment" a non-voluntary, therefore truly political association insofar as one can refuse to enter it only at the total risk of being killed, by enemies in the state of nature, or by the state itself if he or she tries to opt out of the commonwealth, which is what both Hobbes and Rousseau deem to be right.[10]

Interstate war brings us closer to another fundamental concept, *anarchy*, which is to Hobbes the lack of any "common power," his favorite term, while Locke speaks of "common superior" (Locke 1690, chapter 3, 126). Henceforth, I shall use *anarchy*, if not otherwise indicated, with this last and fundamental meaning: a pattern in which, whatever network of partial cooperation and conflict resolution the ties existing between sovereign states as well as international institutions may have developed, if it comes to a clash between the political will of an interstate actor (a single state, or an alliance) and that of another actor, it is at any time possible, though not inevitable, to resort to an armed confrontation, in which the decision as to whose will shall prevail is made. In a however modified Hobbesian framework of sovereign states, the Clausewitzian type of war remains the last resort.

On these premises, we can now tackle the fundamental link between *external and internal sovereignty*. Far from merely meaning independence, sovereignty means a twofold protection: protection against domestic disorder and civil war through the monopoly of force, and enhancement of this centralized force in order to successfully protect the commonwealth against external dangers; but the first condition of external protection is provided by the elimination of the Behemoth of internal disruption, which can make a country the easy prey of another.

What could put an end to anarchy? Evidently only a superior power or third party. But it should be a *tertius super* (not merely an *inter* or *iuxta*) *partes*. That means an actor which

i. is by its own nature obliged to intervene in case of conflict and
ii. whatever the criterion (its pleasure, natural law, international law, ethics of justice, prudential wisdom) it applies to settle or to prevent the conflict, has the means, the "sword," and the determination to enforce its decision. The choice of criteria is indeed not as wide as I have indicated, because some of them are hardly consistent with the legitimacy claims that the "common superior" must be able to successfully sustain. But in principle even a despotic *tertius*, acting at its own pleasure, is an alternative to anarchy, a chance of peace contrary to the resort to war.[11]

A further notion needs to be clarified before we leave Hobbesian ground: the *security dilemma*, now a basic concept of International Relations. It is not easy to establish a common denominator from among the various versions of this term,[12] nonetheless, the dilemma seems fundamentally to lie in the aggravated insecurity which the formation of Leviathans for the sake of individual security and their subsequent power struggles can give rise to. Martin Wight adds to the dilemma a "Hobbesian paradox, namely that the social contract may throw up a tyrant worse than the state of nature" (Wight 1991, 26).

Now, a dilemma is defined by the Random House Dictionary as a situation requiring a choice between equally undesirable alternatives. The problem with "Hobbes' dilemma" is that there are no alternatives: it is not possible for the individuals to choose not to create a commonwealth, out of the preoccupation over subsequent conflicts among commonwealths. This would mean maintaining among individuals conditions that are typical of the state of nature, conditions in which insecurity is incomparably greater than in interstate wars, namely, total. Human individuals *must* seek protection under a common power, otherwise their entire kind is in danger, and the commonwealth in which they live and look after their "industry" will seek by means of war or deterrence (or alliance) to come to terms with the challenge raised by the power of other commonwealths; if it is unsuccessful and is dissolved, its subjects are not as endangered as the commonwealth itself (they do not face *equally* undesirable alternatives), because they still have a get-out clause insofar as they can bow down to the victor. A true "security dilemma" applies, as I shall argue, only to nuclear conditions. In the non-nuclear case it would be more precise to speak of contradiction rather than dilemma with regard to security; but I am not going to propose to change the term, as "security dilemma" is an entrenched notion. In any case, the "dilemma" is no more Hobbesian than it is Lockean or Rousseauian: Locke even devised for his liberal model of government a special federative power, beyond the legislative and executive, to provide for the external security of the commonwealth, which in the state of nature is in its relationship to the others (cf. Locke 1690, chapter 12, 190–92).

As for the "Hobbesian paradox" mentioned by Wight, the problems with Hobbes' absolutist view of state power should be discussed as rigorously as Hobbes formulated them, that is, by taking his concepts (sovereign, state of nature) as political abstractions, with no reference to the individual psychology of the office holder. His sovereign, who is not necessarily a monarch, but could also be an assembly, can only be renamed a "despot" by overstretching Hobbes' own comparison of despotic and sovereign power.[13] However despotic or tyrannical (as he would have said, according to classical terminology) or merely absolute the sovereign may be, it is not a matter of hope (or, as Wight suggested, of Hobbes' inconsistency with his own anthropology) if he or she is less bad and less worthy of distrust than ordinary people. To be the sovereign of the Hobbesian state, a person need not be different from ordinary people, sovereignty being an institution of power, not of morality or charity. As for the distrust, this will depend on his or her performance in ensuring and fostering the *salus populi* (not pure survival, but also wealth through industry) against both the Behemoth of civil disorder and the other Leviathans.

I would now reformulate the philosophical quintessence of the previous, very partial analysis of Hobbes' writings, the "Hobbesian moment," as the basic anthropological and political rule that, whenever life is at stake, and whatever the (domestic or interstate) origin of the danger may be, individuals seek and actually receive protection under a common power, which provides the necessary order by creating peace internally and furnishing adequate defense against enemies from without. Only underneath this protective layer can risks be taken, and non-violent (legal) regulations of conflict-laden matters be developed.

3. ANARCHY, SOVEREIGNTY, SPACE, AND WAR

Far from presenting a summary of modern history, let us now simply ask if this conceptual picture finds correspondence in the political reality during the time in which Hobbes formulated his thoughts about the state and in later times, up to 1914.

The main question is: was it anarchy? It was, in the strictest sense of anarchy as a "lack of regulation through a superior authority." It was not if we think of anarchy as the absence of any regularity in behavior, that is, of any empirical or moral or legal rule, in a word, chaos or "immature anarchy," as Buzan terms it (cf. Buzan 1991, 175). It was not because, in spite of the formal reaffirmation of the authority of the Holy Roman Emperor, the Westphalia peace treaties established a system of states[14] in which some elements of order always existed. It was the "anarchical society" depicted in Hedley Bull's fitting oxymoron,[15] in which "order" means a pattern in domestic or international relations that ensures security, insomuch as it fulfils elementary goals such as those of preserving life, restricting the spread of violence, and stabilizing expectations, because normally *pacta sunt servata* and not simply *servanda*,[16] and possession is respected, or the contrary punished. The guarantee of the preservation of life is applied to all citizens, except when they are appointed to the armed defense of the state against enemies from within or without, or become unavoidably involved in fighting aimed at that defense.

But as *inter-national*, or rather inter-state order, it was and is not based on a world government, but rather on rules of reciprocal behavior, which states mostly apply (though not everyone at the same time), because they mostly share certain goals and a certain framework of action.[17]

Insofar as states share certain general goals, which they regard as the only viable ways to preserve the elementary features of order mentioned above, they do not merely form a system, but an international society or, as commonly known today, community. This exists inasmuch as they are aimed at preserving

i. the society itself and its rules (against actors seeking supremacy and conquest such as the France of the *Roi Soleil* or Napoleon, or Hitler's Germany);
ii. the external sovereignty of any single state (with the exception of Poland or the Central and Southern Italian states during the *Risorgimento*);
iii. international peace in its link with security, thus resorting to war when the balance of power or, later on, the collective security system is endangered.

The framework of action for pursuing those goals is defined by *sovereignty*: based on the internal monopoly of legitimate force in the hands of a single (however inwardly segmented) center of political will, external sovereignty means formal independence from other powers as well as a certain degree of autonomy or effective self-rule. "A certain degree" implies that on the one hand hardly anybody, not even a superpower, can shape its destiny without paying any tribute to external conditioning; on the other hand, below that degree a state has "limited sovereignty" or is a "satellite," as happened in the Soviet bloc from Stalin to Brezhnev and beyond; but examples could be found in the Western Hemisphere during the Cold War as well.

Sovereignty is a Janus-faced notion: it gives a name to the separation of political units, which, "because of their Independency are in continuall jealousies" (Hobbes 1651, part I, chapter 13, 187), are ready for political hostility and war, as soon as they perceive another state's behavior or assumed behavior as an infringement of their formal independence and/or autonomy.[18] On the other hand, sovereignty is not just another name for "jealousy"; it is a political and legal concept, which entails some rational discourse about states in peace and war. As a legal concept, it implies recognition: not merely legal-diplomatic recognition, but the admission of an existing, even if rudimentary, commonality, since recognition is related to reciprocity and equality. As states assert their sovereignty, they want other states to recognize and respect it; the state which asserts its sovereignty thus grants the same status to the states whose recognition it is seeking. As to the equality implied in the notion of sovereignty, it is certainly a *formal* equality, which may coexist with huge in-equality in power and wealth.[19]

The system of sovereign states, and the elements of society which developed within it, cannot however be properly understood without taking into account one of their preconditions: the new *spatial order of civilization and politics*.

First, sovereignty is not conceivable if the central power is not protected against the ruinous consequences of political or military defeats by a "hard shell" (cf. Herz 1959, part 1) surrounded by enough territory as to create a "fortified buffer zone" (Tilly 1990, 70). Conversely, only the clear-cut delim-itation of the territory over which sovereignty is exerted and the definition of borders enabled the creation of a framework in which institutions like *cuius regio, eius religio*, or the transformation of the *ius gentium* into a *ius inter gentes* (sc. *Europaeas*), implying respect for the domestic jurisdiction of each state, could make sense. The creation or consolidation of large territorial states (*Flächenstaaten*) was the geopolitical premise of *ius publicum eu-ropaeum*, with its regulation of violence.[20] But equally important was the dis-covery, occupation, and division (*Landnahme*) of extra-European continents, especially America: colonial compensation became a tool of European diplo-macy, while the legitimacy of ruthless war and privateering beyond the "amity lines," agreed upon by the European powers,[21] helped to reduce the amount of violence used on the old continent. It is true that the origin of a structure or institution, which may evolve and adapt, is not *per se* a judgment of its validity. But if we look at the total change in the spatial order of poli-tics, up to the military activities carried out in outer space, we can intuitively realize that the patterns set up in early modernity, like sovereign statehood and the democratic regulation of power in it, need thorough theoretical revi-sion. This ridicules all pretensions to come to terms with the contemporary world using exactly the same conceptual tools (national sovereignty, national interest, and national security) that were designed in the eighteenth century.

In this spatial framework, balance of power and war were the leading pat-terns for the self-regulation of international society.[22] In fact, the common in-terest in preventing one of them gaining supremacy was explicitly acknowl-edged by the European powers in the peace treaty of Utrecht in 1713. What in other times and regions may have been a hidden pattern of international be-havior became conscious policy (Herz 1959, 65), guaranteed by the well-protected insular nature of the guardian of the balance, Britain, as well as a diplomacy method which handled interstate relations as a pure matter of ter-ritorial power (ibidem, 62–68). The very nature of the system as well as its publicized (most eloquently by Vattel) legitimacy was the basis on which it exerted political deterrence against any supremacy-seeking country.[23] Were political deterrence to fail, so the other regulator, war, had to intervene: by denial in Europe, stopping territorial gains, and/or by punishment in the colonies of the "aggressor." But it was war between equal sovereign states, fought by armies (often mercenaries) regarded as the executive arm of governments which would thus decide whose political will had to prevail; not war against the population, nor for the triumph of the true faith or the just (moral, ideological) cause. It was rather Schmitt's *nichtdiskriminierender Staatenkrieg*, the war among enemies who do not claim any moral or religious superiority over each other and do not discriminate between just and unjust warring states. In this sense, the medieval notion of just war became obsolete, but the just war tradition made an important contribution to the "bracketing of war" (Schmitt's *Hegung des Krieges*): it underwent a process of secularization and proceduralization. A war was regarded as just if fought, after due declaration, between sovereigns which were *iustus hostis* for each other, each of them entitled to choose this way to resolve their controversy, war being their last resort and peace the ultimate aim of their actions.[24]

It must be clear that the cultural and juridical restraints of war could be as effective as they actually were only because the framework of technology, geopolitics, and power distribution shaped war (the *is* and the *ought* of war) as a substantially political action. Nobody has had a sharper cognition of the *political, and therefore ambivalent* nature of modern war than Carl von Clausewitz. Both versions of war are political: the clash between the wills of states is political, as they attempt to impose their will on each other, ready to devastate the adversary until it yields ("absolute war"). Even more evidently political is actual moderation in warfare, induced by the circumstances in the behavior of the parties, thus giving war its effective shape ("real war, *wirklicher Krieg*"; Clausewitz 1832, 579–81). It was the strictly political view of war as a "continuation of policy by other means" (Clausewitz 1832, 87)[25] that restrained war with reference to the original ends of politics such as the self-affirmation and preservation of state power in an environment defined by the existence of other states. But things are different whenever politics is subordinated to other ends such as the triumph of a civilization or religion or ideology.

War has indeed been pivotal in the making (and later unmaking) of modern Europe (cf. Howard 1984), both in terms of actual war and war preparedness. The typical European state was thus shaped not simply by the monopoly of concentrated coercive force of which it was supposed to consist; similarly important was the build-up of the bureaucratic (financial, technical) machinery that was required to extract the means necessary for warfare from the land and peoples.[26] The outcome of the establishment of Leviathans in Europe was "less fear and more security," not merely in the picture sketched by philosophers, but in the life of the people. In wars involving great powers, battle deaths diminished greatly from the seventeenth to the eighteenth century, and dropped sharply, the Napoleonic wars notwithstanding, in the nineteenth century, with numbers skyrocketing again in the last century. In the same way, the number of years in which great powers were at war dropped from 94 in the seventeenth to 40 in the nineteenth century, although the total number of wars, including civil wars, rose from 50 in the eighteenth to 208 in the nineteenth century (Tilly 1990, 72–73). Perhaps with the exception of Napoleon's campaign in Russia, no European war between 1648 and 1914 had the same disruptive effects on the population as the Thirty Years' War, in which the warring parties still acted as anarchical Leviathans.[27]

Protection against invasion, occupation, and plundering from the outside was intertwined with the domestic monopoly of coercion and led to an overall decrease in violence. Once the privileges and widespread violence exerted by the aristocracy had been successfully combated,[28] urban violence became a government issue, pursued with a varying rate of success, but with the overall trend of at least lowering the amount of violence against persons (Delumeau 1989, 554), while enhancing and rationalizing the state's punitive violence against outlaws and social rebels, as Marx and Foucault have shown in their inquiries. The increasing domestic order created the conditions under which other sources of insecurity could be addressed by both free enterprise and public administration: the creation of professional firefighting, the improvement of public health, and the expansion of the insurance business to more and more branches of property and activity carry evidence of this link. Under Leviathan an ordered, though unequal civil society did develop, a further key element of modern rationality.

4. FEAR, INSTITUTIONS, AND INTERNATIONAL SOCIETY

Before we examine the last of these elements, international law, in § 5, we need to refine two categories of fundamental importance not just for this chapter: fear and institution.

Fear retreated as the leading motivation behind state-building, which was insofar successful. Of the state functions, war-making and war-preparation grew at a slower pace than those connected to the implementation and reform of the legal system and the regulation of the production and distribution of goods and services (cf. Tilly 1990, 97). To a certain extent, protection against crude, extreme insecurity became usual, and the expectations related to it could be stabilized. In other words, fear and insecurity became neutralized and depoliticized, even though this could be reversed.[29] Does this mean that fear and security are transient elements of politics, peculiar to the historical situation of the sixteenth and seventeenth centuries alone?

This would be a historicist reduction of theoretical and structural problems. It does not take the possibility of nuclear holocaust to be aware of the *permanence of fear* in the background of our social life; a background from which it can be rapidly recalled and reactivated, as soon as war or terrorism or civil disorder or dramatic impoverishment surface. This is a complex mechanism, but let us take fear into account only to the extent it directly impinges upon politics, particularly in the build-up and disruption of political institutions. Fear of death for example has an influence on how we lead our individual life, including participating in public life or abstaining from it; but what matters directly in politics is the fear of being killed by factors which are in themselves political (war, revolution, tyranny) or politically controllable, which is now also the case for many citizens with the greenhouse gas emissions that cause global warming. Nowadays the list of feared factors which can be put under political control is not the same as it was 200 or 100 years ago: who would have imagined then that the preoccupation of possibly being killed in a work-related accident or by a treatable, but in the current health system uncured, disease would become a political matter? This is an issue for chapter seven, but here it lets us note that the relevance of fear in politics is met neither by the minimizing, historicist view we have mentioned, nor by the *topos* of political realism which states that "all politics is regulation of fear," as it claims to provide a universal key for all political contexts, regardless of historical change and conceptual restructuring.

A last clarification regarding fear is concerned with its psychological or psychoanalytical background, which often re-echoes in its political understanding. A first example: the distinction (see also Delumeau 1989, 30) between fear (*peur, crainte, terreur*) as related to what is known, and anxiety (*angoisse, anxiété*) as related to the unknown is a widespread notion, but clear and general criteria which might uphold that distinction are not provided. Appearing more convincing is the Freudian distinction, applied by Franz Neumann in his landmark essay of 1954, between neurotic fear (*neurotische, automatische Angst*) and fear before a real danger (*Realangst*), the former being a matter for authoritarian manipulation, the latter the basis of political association. The problem is that, while for the individual patient an external agency (the society, the analyst) has the authority to say which kind of fear prevails in a single situation, no similar institution exists in the case of a large group like a nation (cf. Neumann 1957). This shows that the political problem of fear cannot be addressed from a psychic point of view alone, because politics is made by individuals driven by psychosocial factors as well as by institutions that contribute to orientating and restraining the actions of the individuals and to defining their meaning.[30] It therefore seems reasonable to conclude this terminological detour here.

Let us pass over to the other category. In social and political science, *institutions* are generally understood to be "persistent and connected sets of rules (formal or informal) that prescribe behavioral roles, constrain activity, and shape expectations" (Keohane 1989, 163).[31] Like the rules they refer to, norms and sanctions can be formal and even legal, or informal, that is, consisting of customary patterns of behavior. "Institution" should not be merged with whatever regularity might be observed in the behavior of social actors, for example with the sociological rules or "generalizations from experience" mentioned by Max Weber (Weber 1922, 11). In other words, a mere rule is a rule for the observer, not the participant; while an institution is an institution for the participant, as its rules are perceived as meaningful with respect to his or her cultural values and can be internalized. This restriction is necessary for this notion to be used consistently in political theory; it differentiates our use from the anthropological meaning of "institution" as adopted in Gehlen's theory. When applied to international relations, the (restricted) definition of institution includes, according to Keohane (1989, 3–4), "formal intergovernmental or cross-national nongovernmental organizations" as well as international regimes and conventions (like diplomatic immunity before its codification). In a society like the international one, in which no common power exists, the existence of an institution, even if not highly formalized, guarantees the effectiveness (spontaneous or enforced) of the intended rules (cf. Keohane 1989, 7). However, a minimal degree of effectiveness is required, at least in the sense that sooner or later behavior contrary to that which is expected according to institutional rules (whether it be the infringement of rules or the failure to implement them) will be regarded as illegitimate and will damage the image and interests of the perpetrator. Last but not least, institutions are effective because they do not just enforce rules by sanctions, but predefine the interpretation of rules by the actors and shape their expectations.[32]

What we have so far looked at as the emergence of an *international society* in Europe between 1648 and 1914 can now be conceptualized in terms of institutions[33] which do not undo anarchy, but lead it to tensely coexist with certain elements of order.[34] That anarchy was not undone, means primarily that war remained the final and principal regulator between states with conflicting wills. But war, as long as it was a non-discriminating and "bracketed" war between states, was at the same time an institution supporting that order; or rather, it was an accepted rule to resort to this kind of war in order to keep or restore the balance of power, or later to complete state-building for a "delayed" country, like Germany or Italy.

It is finally noteworthy that the way that war was effectively shaped and regulated did not change with the constitutional and later liberal-democratic transformation of European states; on the contrary, *popular* sovereignty, based as it was on the merging of democratic *demos* with national *etnos*, brought ideologization and emotional overload to warfare, even though this remained "bracketed" until 1914. Kant's hope that permanent peace would result from a "regime change" (in current times it is wise to add to this expression: *absit iniuria verbo*) towards a "republican" structure of the former absolutist state, based on the citizens' freedom and equality, remained unfulfilled.

5. INTERNATIONAL LAW

A significant institution of the rational modern order was and still is international law, a term first introduced by Bentham as he argued that the *ius gentium* is nothing else than *ius inter gentes* (cf. Nardin 1983, 118n.). With this change in terminology, the particular nature of international law was acknowledged as early as the eighteenth century: a law having a large body and including several matters,[35] but not issued or enforced or interpreted by an authoritative source, and therefore based on customs and treaties or the "law of nature." In this case international law tended to be doctrine rather than law. It was only after this system had fallen apart in World War I that an organ entitled to interpret international law, the Permanent Court of International Justice, was created. Its history, including its present successor, the International Court of Justice, shows both the great novelty and the grave limits of international law, which can lead to adjudications that no common power can however enforce. It is too early to make any assessment of the International Criminal Court, which has just started working despite opposition from the United States and other powers such as Russia, China, the Koreas, and Zimbabwe, and is expected to systemize the activity of the *ad hoc* courts on the crimes committed in ex-Yugoslavia and Rwanda.

Whatever the present state of affairs in international law, to which I shall come back in chapter seven, within the range of what we could call "classical modernity" (1648–1914) I share Nardin's picture of the European society of states as defined not by "their geographic propinquity, frequent transactions, or common cultural heritage, but rather (by) the fact that they were associated within the framework of a single body of international law" (Nardin 1983, 66).

This *ius publicum europaeum*[36] did not stop war or conquest, or other major violations of its substantive rules. Within these limits, the establishment and implementation of substantive norms was indeed less important than the creation and development of procedures. Procedures like the diffusion of embassies and consulates, the multiplication of treaties, conventions, and agreements and holding international conferences (beginning with Vienna 1814, or perhaps we can go back to Westphalia 1648) created the framework in which states would be able to seek peaceful compromises in case of conflicting interests, or regulations to prevent the upsurge of conflicts altogether.

The *proceduralization* of relationships and conflicts can be seen as a Grotian rationalization of international relations as relations among states, the ultimate and sovereign subjects of all processes taking place in the arena.[37] Proceduralizing the relationships between the Leviathans gives them an element of rationality in both aspects of this notion. It means first that their self-affirmation is no longer necessarily sought in agonistic relations, or zero-sum games; these remain possible as a last resort, but in the first instance the way to peaceful compromise, or positive-sum games, is systematically and permanently open, according to existent and available procedures. These procedures can be reactivated even in the middle of a military campaign: *silent inter arma leges* is no longer completely true, laws of war were established and from time to time observed, even the law we are hinting at is a *sui generis* law. What does all this mean?

Nardin has convincingly argued that international law *is* law, because the absence of a legislative organ as well as the usual forms of enforcement and authoritative determination do not debase its nature of a system of rules, whose criteria of validity are internal to the system itself (Nardin 1983, part two). I have added the *sui generis* clause because on the one hand I share his view of international law being law and reject any legal version of the domestic analogy (rules are law only to the extent they are like the enforced rules of the legal system within the modern state). On the other hand, the stress on the particular and limited nature of this type of law is necessary, because law cannot be dissociated from the *effectiveness problem*, as if the self-contained nature of law could shield it from any checks under the "principle of reality." In certain fields of ultimate power games, and under certain circumstances, not just the rules or doctrines, but the very adjudications pronounced by the International Court of Justice have been as ineffective as to deprive the law of its basic quality of being the protection of rights and an alternative to violence.[38] Certainly, the law remains law even after proving its ineffectiveness: but in the life of citizens and countries it may then lack credibility and relevance to such an extent as to be threatened by futility; even more so if compared with the tragic matters of life and death that it turns out to be unable to have an impact upon.

Under these circumstances, international law looks like *habeas corpus* or free speech in a country of which several regions are under tyrannical rule: one can struggle by (violent or non-violent) political means to reassert them, but will hardly behave as if legal norms and procedures should fully shape and restrain one's own conduct before the rule of law is effectively reestablished. Now, in our civilization tyranny is fought against as a temporary state of affairs: once the tyrant and his or her regime is toppled, the law can be expected to rule again, although this expectation has at times been betrayed. In international affairs, even when a blatant violation of the law was reversed and punished, it rarely happened as an automatic, reliable performance of the legal system; this lacks, among others, one fundamental feature of the domestic protection of rights and enforcement of order: it does not activate itself, but is only activated through political decision (generally of the UN Security Council), a circumstance which makes the mechanism volatile and precarious.

My attempt to understand the real rather than the declared way in which the international legal system, as an independent and *sui generis* entity vis-à-vis the domestic one, works, is a prelude to asking, in later chapters, if and how far it can change and is actually changing once confronted with new requirements and trends. It is not true that, once questions of life and death arise, international law never plays a role. We have at least some evidence of the contrary, even in the nuclear era and in the most risky (and paradigmatic) crisis of this era, the Cuban missile crisis of 1962, as in its multifaceted functions the system was able to put some legal constraints on the decisions to be made, then pushing, according to Abram Chayes' account, the political decision-makers to look for a legal justification for the option they were oriented to choose, and providing an institutional frame (the Organization of the American States and the United Nations) for the implementation of these processes.[39] While Chayes himself wisely marks the limits of international law vis-à-vis the centrality of political process, I would like to add that even this limited, but effective impact of international law on the outcome of a high-stakes conflict was, as the case of the Cuban crisis showed, made possible by a set of (fortunate) personal and structural circumstances, whose presence in other cases is highly uncertain.[40]

This limited, but effective impact of the *ius inter gentes* may have been satisfactory in a world in which, between 1648 and 1914,

a. no global, ultimate challenges existed,
b. a more comprehensive and effective solution was hardly available, and
c. the attempts made to attain it were not only unsuccessful, but, as Nardin suggests, may have damaged the effectiveness of the existing legal norms and institutions. Whenever *ultimate conditions* of societal life are at stake, it is not just the limited amount of safety provided by international law, but its uncertainty and precariousness that raise the question of whether a dif-

ferent type of law, as well as a different link between law and politics in the worldwide dimension, is not merely desirable, but necessary.

6. RATIONALITY, ONE or TWO?

To the critical reader the structures and institutions of political rationality which we have so far given account of may seem to be nothing but the strategic rationality of an enlightened Hobbesian sovereign: the very first rule of the law of nature, self-preservation, makes states seek not a common power, but different, not necessarily violent ways of pursuing their interests; the resort to violence remains in any case available. This procedural rationality can result in fostering peaceful relations, insofar as they do not exceed the status quo as determined by the actors' actual preferences; these do not allow in general for the change requested by new emerging actors and/or preferences, for example, to enhance the importance of liberty and/or equality, or human rights. In the worst case, this kind of "neutral" rationality is flexible enough as to possibly provide a legal form and justification for situations resulting from acts of violence or injustice inflicted on sub-actors, like minorities, or not-yet-actors, like colonial or irredentist peoples. Rationality can purely mean—our critic would conclude—the rationalization of acts as well as their outcomes, with no possibility of questioning their compatibility with Reason as the directive force and chief goal of humanity.

The philosophy and ethics of international relations indeed provide a good vantage point for watching the dialectics of the different models of rationality that have shaped modern thought about politics, society, and history. However different those models might have been, they can be referred—simplification is lamentable, but necessary—to a fundamental dichotomy. On the one hand, we find the rationality which we have so far termed *strategic* (in a sociological, not military sense!): it tends to improve the effectiveness and efficiency[41] of a given structure or system. In doing so, this model of rationality does not raise questions as to whether the system's existence and goals can be justified according to general values and interpretations; moreover, it can even contest the sense and admissibility of those questions, claiming that only strategic or technical rationality deserves this name. The rationality envisaged by neoclassical economics, functionalist sociology, and to an extent the system theory of society can be referred to in this model.

In the other model, rationality is regarded as being embodied in a set of fundamental values (first of all, autonomy of the individual) and principles (peace, liberty, justice, and equality) that focus on the elimination of suffering and improving the well-being of human, rational beings living in society.[42]

This model of reason is not analytical and instrumental, but rather value-laden and normative: it entails the idea that reality, as it is, does not correspond to the values set by reason, and needs to be changed. The project to change it may or may not be embedded in a philosophy of history providing a general explanation of its course and forces as well as predictions about the change to come.[43] Vis-à-vis the strategic model, we can call this one *substantive*, even though I do not forget that not every substantive notion of reason envisages the criticism and renewal of reality; for example, Hegel's reason only half-heartedly did so, except in the interpretation of the Hegelian Left.

These are the two somewhat simplified models of reason in modernity. Unlike Weber's "ideal types," which by definition never come up in an undiluted, exclusive version, our models may, though rarely, do so. Strategic rationality may attempt to rule out the very legitimacy of substantive reason, the latter may presume to be able not merely to subdue the former, but rather to overrun its specific requirements altogether. Their complex relationship cannot be examined here,[44] nor can we widely discuss what degree of correspondence may exist between this dichotomy and the dichotomy in doctrines of international politics that now draws our attention. I am less interested here in this dichotomy as irreconcilable opposition than in a dichotomic distinction which may allow for intertwinement and partial continuity between the two forms of rational discourse about states and peoples. This interest is related to my thesis, which shall be discussed in the last section of chapter three, that both kinds of rationality have been deeply affected in the twentieth century, most dramatically by the rise of global challenges.

Now, since the very beginning, the rationalization of international relations, a major step in the modernization process, has been matched by protest in the name of a different idea of reason.

"Ah philosophe barbare, viens nous lire ton livre sur le champ de bataille!"

In Rousseau's invective at the end of his horrifying description (Rousseau 1755–56, 43) of a battlefield, on which a barbarous philosopher is sarcastically invited to read from his book, it is not Hobbes alone who is meant, but rather every philosopher who tries to find some rationality in war and violence. The attitude exemplified by this quotation generally claims a totally different kind of relationship between peoples as well as individuals, a relationship shaped by reason as a benevolent force which lets all inhabitants of the planet finally recognize the rightness and necessity of conciliation and co-operation. The most widespread version of this kind of rationality is the Kantian idea that politics, including international politics, should and indeed can be put under the mastery of law; instead of the law of treaties, the universal law of Reason, whose essential aim is to root out violence and let peace and justice govern all our relations.[45] On this path, Reason is often accompanied or superseded by other forces, like love or imagination, and the rational design of the future contains an element of utopia (in the sense of what morally ought to be and also can be attained, but is not yet real). Pacifist or religious cosmopolitanism, from William Penn to Charles Fourier, as well as the radical liberalism of Norman Angell's *The Great Illusion* (in the first edition of 1909), may be regarded as modern examples of this attitude. Religious cosmopolitanism is not necessarily progressive, as the Holy Alliance of 1815, or rather its ideological documents, remind us (cf. Nardin 1983, 91–92). Romantic antirationalism and antimodernism can be another outcome of the radical refusal of established political rationality, as the classic example of Novalis' *Die Christenheit oder Europa* illustrates.

The second type of reaction to strategic rationalization in modern international relations is quite different from, though not opposite to, the first one: it is the idea of reforming international relations, developing and improving their rationality in a way which is expected to bring us closer to the goals of Reason (peace and rule of law). But this is done by maintaining states and their society as legitimate and perfectible actors, thus accepting the possibility that the final goals, as well as those closer to us, may never be attained or may be put outside the reach of the present generations. This is in a narrow sense the internationalist way to rationalization; the Grotian path, according to the trichotomy designed by Wight (1991). This "reformist" attitude can be tied in various ways to the radical one: Rousseau's pacifist invective in *The State of War* is not a refusal of any rationalization, and is framed by his attempts to design a different *ratio* of international relations and war as mere interstate relationships, not affecting the citizens as such—the position which was a few years later formulated in *The Social Contract*. Kant's basic view goes a step further than Rousseau's: war as interstate war is legitimate only to prevent aggression and, as preemptive war, to keep the balance of power (Kant 1797, §§ 53–61, 150–57), never as punitive or occupational war. Kant occasionally indeed went so far, as in § 28 of the *Critique of Judgment*, as to appraise warfare's culturally "sublime" features, provided the laws of war are respected. But his internationalist rationalization of war then develops into the normative setting of substantive goals of Reason: from the very idea of law he derives the radical ideal of eternal peace through a permanent congress of states; an unattainable, but practically and legally influential goal (1797, §§ 61–62, 158–59). Furthermore, he designs a further dimension of law, going beyond the law of peoples (*Völkerrecht*) as the law of states: cosmopolitical law defines our rights as dwellers on the earth as valid against all states, that is, our rights as world citizens (*Weltbürgerrecht*). A few years before, in

Perpetual Peace (1795), he had asserted the superiority of a liberal, if unattainable world state (a "world republic"), and put his entire view of domestic and international politics under both the imperative and the asserted possibility of its complete moralization.

Kant is perhaps the best example of the internationalist position being distinct from, but not incompatible with the universalistic one.[46] This attitude is better described as radical reformism: it amounts to the perspective of a world freed from war and violent domination as an effect of (domestic) social and political struggle. Already in the early seventeenth century the European structure of sovereign states and interstate wars, which was about to come into being, was regarded by Eméric Crucé as flawed by international misunderstandings as well as the dominance of a warrior class, which were both to be outplayed by the expansion of free trade (see Howard 1978, 20). The liberal *topos* of economic and cultural transformation through free trade, as well as the idea of political reform, are the elements of Montesquieu's vision: "The spirit of monarchy is war and expansion, the spirit of republics is peace and moderation" (Montesquieu 1748, 132).

The second of these elements, republicanism, played a major role in Kant's view of the interaction of domestic and international politics. A few decades later, two other radical reformers, Marx and Engels, were convinced that eliminating the mechanism of capitalist exploitation and unifying the world's working class would uproot the very source of state violence and war. At the end of the nineteenth century, Marxist theory and the socialist movement were clearly aware of the link between increases in exploitation and increases in military expenditures.[47] But Marxism's radical universalism, which was unable to see that even a socialist state remains a Leviathan, was connected to a peculiar and, as it later turned out, fatal kind of political realism: in order to achieve peace and liberation, the world has to accept the necessity to go through decades of "wars of peoples and classes," as Marx and Engels (a military expert whose nickname was *der General*) said in the early 1850s, as they assailed the pacifism of "utopian socialists."

The connection of "proletarian internationalism," born as a vision of peace and brotherhood,[48] and a realism focused on military power, up to the ideologically "just war" in defense of socialism, is remembered here only to illustrate a case of a merger which is far from rare and can reap unexpected results. Similarly, the three critical approaches to modern political rationality in interstate relations may have come up at times in their distinctive form, but they have also been, and not less frequently, interlinked. Of these, the cases in which the internationalist and universalistic approaches merged have been particularly productive. In a kind of reciprocal self-criticism, the latter worked as permanent stimulus to the former as well as a measure of its achievements, or as a "regulative idea," as Kant would have said. On its part, the internationalist approach helped to keep the universalistic project within the political realm, preserving it from sliding into sectarian velleity. Law, including international law, stays on the border between the two poles of rationality, and represents both their possible, intimate connection and their difference; a difference which is mirrored within law itself as a difference between international law and (so far merely doctrinal) world law.

I assume that this interaction occurred, when and as far as it occurred, because between states a fairly stable environment of *moderate challenges and stakes* existed: with the few exceptions noted above, war was never a "deadly quarrel" to a country, let alone humankind, and some kind of acceptable balance could in any case be reestablished.[49] Nor were the unintended effects of technology a planetary threat, as in the case of global warming.

This balanced structure of interstate affairs was significant for the domestic domain too. It was in this "realm of relative reason" (Schmitt 1950, 142) that some rationality of history could be assumed as the condition that makes philosophical and political projects of progress (Condorcet, Saint Simon, Comte), spiritual improvement (Hegel), or emancipation (Marxist and non-Marxist socialism) possible. Even a philosopher who was far from any historicist approach like Kant could not exclude "nature's secret plan" (*geheime Naturabsicht*) supporting the progress of humankind towards cosmopolitanism, as he did in his *Idea for a Universal History from a Cosmopolitan Point of View* from 1784 (Kant 1988, 422). External stability and moderate challenges did in fact favor the implementation of major political ideas like popular sovereignty and democracy: between 1812 and 1917 the United States only fought marginal international wars (the Mexican and Indian Wars, the war with Spain in 1898) and between 1815 and 1914 so did Britain, since the war in Crimea as well as the colonial wars in Africa and Asia never put Britain's power (as in the Great War) or existence (as in the Battle of Britain of 1940) at stake; nor did she experience any civil war in that century. It seems that, for the democratic regime to put down roots in a country, some measure of neutralization and depoliticization of fear and insecurity is needed. Only then do (national) democratic processes make sense and earn the people's loyalty, since it becomes evident to everybody that democracy really has a positive impact on the people's destiny.[50] Current theories of democracy still struggle to reflect in conceptual terms this intertwinement of domestic and international aspects.

As a conclusion to this chapter, I want to clarify some conceptual questions underlying the categories used in it. Its main topic, modern political rationality, is not the sum of the processes that are commonly described as "modernization": more innovation, more differentiation, more efficiency. Nor is the rationality of modernity limited to its anti-traditionalistic and disenchanting impulse, whose classical account goes back to Max Weber. In the language used in this book the rationality of modernity is a multilayered concept:

i.   It means the *ratio essendi* of political modernity as a structure developing around some basic features (sovereign states + "bracketed war" + balance of power, all wrapped in a more or less effective legal regulation).
ii. It means the claim of actors and interpreters to be identified as modern (reflexive modernity) insofar as they are allegedly part of, or argue the case for a rational order of political life; an order which includes international affairs, which are its less rationalizable part. As a general claim, it has been made by philosophers as different as Hobbes, Voltaire, and Hegel, as well as by international lawyers such as Grotius and Vattel.
iii. Finally, it means our judgment, as observers as well as the skeptical heirs of those actors, that, within the structure indicated under i., and as the outcome of the claims and projects ("the project of modernity," as Habermas put it) indicated under ii., something was attained which partially and temporarily *fulfilled the claim of heading for greater rationality* made by modern politics. Something which does not satisfy our (normative) idea of rationality and reason, but at least goes in the direction of what we mean by these words. I do not refer to their meanings in the several philosophical debates on reason, but rather in common language, the mirror of our "life world": what can be regarded as the main achievement of modern rationality is the decrease in the destruction, death tolls, and unrest paid by Europeans[51] between 1648 and 1914 than in the time before and after those chronological boundaries. A rationality, for sure, whose relativity is aggravated by the enslavement of millions of Africans and the genocide of the American Indians perpetrated by Europeans and later Americans, as we will see in § 1 of the next chapter, dedicated along with chapter four, to the danger that this story will terminate in a very irrational manner.

NOTES

1. The semantic history of this word in the Western European languages gives interesting clues as to its political developments, cf. Conze 1984 and Delumeau 1989, 9–29. Of the two words into which *securitas* splits, indeed not only in English (security and safety), I shall mostly use security, which is closer to its abstract political meaning.

2. Cf. Gehlen 1980, a translation of *Die Seele im technischen Zeitalter*, while Gehlen's defining book *Der Mensch* is not translated into English. Leaving aside the

complex case of Sartre, we can see in both the cases of Heidegger and Gehlen that their deep sense of human precariousness was not accompanied by a critical analysis of political life as the place in which reassurance can be sought, as Hannah Arendt did for example in her attempt to revive the *polis*. Their link with the Nazi regime was—as we can see in the better known case of Heidegger—not a mere personal episode, but rather a theoretical misstep towards placing in the community of the *Volk* the source of reassurance for the individual about the meaning of his existence, through a renewed link to the Being (*Seinsverbundenheit*).

3. What else is the central theme of *virtù contra fortuna* (in *The Prince*), if not a reformulation of the question of how to make the power and unity of a commonwealth secure against civic disorder and foreign occupation, specifically in a world, like that of the early Renaissance, which had not yet learnt to come to terms with time, change, and upheaval? The motif of virtue which may be able to stabilize the political order against changing circumstances by providing the commonwealth with *buone leggi* and *buone armi* can be found throughout *The Prince*. On time and politics in the political thought of early modernity cf. Pocock 1975. Cesare Borgia's dinner stratagem is narrated in both *Descrizione del modo tenuto dallo Duca Valentino nello ammazzare Vitellozzo Vitelli, Oliverotto da Fermo, il signor Pagolo e il Duca di Gravina Orsini* and in chapter 7 of *The Prince*.

4. The philosophical view that sees security as a fundamental category does not exclude that under specific political constellations security can be used as a label ("securitization") for a presumably existential threat that justifies the use of force, thus favoring conservation of the traditional order, which is what some IR schools of thought in Europe are focused upon, see Waever 2004. Conversely, this kind of critique of ideology should not pretend to abolish or relativize essential notions of political philosophy.

5. "The noble knight" remembered in the songs of the Royal Imperial Army.

6. "Security is an abstract notion which came into being along with the principalities of European modernity. As a concept it was a feature of the new (military) state and derived from its reason... Security became one of the leading concepts in the new European state system and its law of peoples" (Conze 1984, 842, my translation). Security among multiple independent actors (the first definition of sovereignty was given in 1562 by Jean Bodin in *Six livres de la République*) thus replaced peace (*pax, Frieden*) under one (or two, the Papacy and the Holy Roman Empire) universal authority as the main goal of the political framework.

As for the need for security in the everyday life of men and women, and how the new state addressed it in a secularized way (public order policies replaced religious rituals), cf. Delumeau 1989. *Rassurer et protéger vs. surveiller et punir*: Delumeau's focus on how the functions of reassuring and protecting were socially executed at the same time contradicts and complements Foucault's attention on the repressive mechanisms of power.

7. In the Weberian tradition, I keep the empirically verifiable consensus distinct from legitimacy, which is the chance for an institution or policy to be justified on the basis of a meaningful set of fundamental values, beliefs, and worldviews. Without this background it would be impossible to understand consensus. Likewise, it would be false to say that consensus is the mere manifestation of legitimacy.

8. The oppressive accumulation of (bureaucratic) power in the modern state, of which Marx and Engels experienced a backward version in contemporary Prussia as well as in the doctrine of its leading political philosopher, Hegel, was explained with the up-and-coming bourgeoisie's need to secure power. Security is in this sense the paramount category of modern political philosophy, as Marx put it in *The Holy Family*. Marx, however, wanted to preserve centralization as an organizational technique even after the proletarian revolution has dissolved the oppressive state machinery, which is to be replaced by self-government (Marx, *The Eighteenth Brumaire of Louis Bonaparte*, written in 1852, and *The Civil War in France*, written in 1871).

9. For a comparison with Spinoza's similar position in *Tractatus politicus*, cf. Bull 1977, 49.

10. There is, in addition, a parallel meaning of the state as a non-voluntary association: it refers to the destiny of the single person, who becomes citizen of state *x* rather than *y*, or rather than stateless, simply because he or she was born in its territory or from parents having its citizenship.

11. On all the legal and political topics of the third party, cf. Portinaro 1986, particularly chapter 10.

12. Introduced by John Herz in 1950, this notion was later discussed, among others, by Nardin 1983, 38; Lieber 1988, 328; Keohane 1993. In his criticism of Hobbes, Rousseau (1755–56, 45) saw the link between the search for peace and the resulting war, but he did not label it as a dilemma: "Far from the state of war being natural to man, war springs from peace, or at least from the precautions men have taken to ensure a lasting peace."

13. The way in which he uses the term "despoticall" (cf. Hobbes 1651, part II, chapter 20, particularly 255–61) may be misleading; it does merely refer to the power acquired by conquest over a vanquished people, and its affinity with political power simply means that a monarch over diverse nations is sovereign everywhere, whatever (institution or conquest) the source of his or her power. He or she is not a despot in the Aristotelian (a pre-political, private power, chiefly over slaves) or derogative (a bad ruler *e defectu tituli*) sense. What familiar and despotic power means in Hobbes is best explained in Gauthier 1969, 112–19.

14. With regard to the German states after 1648, this term was used by Pufendorf in *De systematibus civitatum*, published in 1675. But the early modern use of this term, which retained many elements of what was later called the "society of states" (cf. Bull 1977, 12) is different from the current signification: I am referring to a set of states among which there is contact as well as a degree of influence on each other.

15. Hereafter this paragraph heavily relies on Bull's terminology; I shall not give account of every single debt, nor discuss what I am going to modify. The distance from his perspective will come up later.

16. In the sense that the respect of the covenants is effectively and not only normatively prescribed.

17. The reader may now start intuitively comparing the following features with those known from the history of the twentieth century and take stock of the changes. An analytical account will follow in chapter three.

18. "It is the characteristic of sovereignty," as Leibniz (quoted in Herz 1959, 57, transl. from the French, mine) put it in *Entretiens de Philarète et d'Eugène sur le droit d'ambassade*, "that it cannot be constrained other than by waging an armed clash."

19. As Christian Wolff (quoted in Nardin 1983, 54) said in *Jus gentium*, a nation, however small, is a nation in the same way as a dwarf is no less a human being than the tallest man; which is not a great advantage to the dwarf, but—to complete the argument—is better than being regarded as a slave or a thing. On equality as an element of the peaceful face of sovereignty and Kelsen's criticism of this notion see Gilson 1984, 59–60.

20. This paragraph relies heavily on Schmitt 1950, particularly on part 3, 140–213; but the considerations developed in this book by the former Nazi jurist are partly coincident with (and cited in) those of the German-Jewish emigrant John Herz, cf.. Herz 1959. I believe that illuminating topics such as the role of the spatial order in politics and law can be taken into account, since they are not inevitably entangled with Schmitt's antirationalist philosophy of the original and determining link between the earth and order. Compare the relevance of a line in space such as the American frontier in the shaping of democracy in the United States, as accounted for in Frederick J. Turner's famous essay (Turner 1893).

21. For whom extra-European units were an object of conquest and/or mission, like the empires of the Aztecs and Incas. The Ottoman Empire itself, partly integrated in the European power system since earlier times (remember its recurrent alliance with France), was officially admitted to the diplomatic "concert of nations" not earlier than 1856.

22. I do not see enough reasons for adding—as Bull 1977 does in the second part of the book—three more pillars of international order (international law, diplomacy, and great powers), which are not as primary as the two major regulators of order and power. Yet the role of international law requires differentiated assessment, which will be done later in this chapter.

23. On prenuclear deterrence see Mearsheimer 1983 (deterrence is broadly defined as "a function of the relationship between the perceived political benefits resulting from military action and a number of nonmilitary as well as military costs and risks," 14).

24. This constituted the *ius ad bellum*; as for the *ius in bello*, the leading standards were the proportionality of good over evil (a principle also influencing the right to go to war) and the discrimination of non-combatants. This summary of just-war thinking in modern times is indebted to Johnson 1980 and Johnson 1984.

25. Even in Hegel's (1821) notorious eulogy of war (§§ 324–40, 360–71) as an expression of the state's ethical substance, war remains substantially a political conflict. Furthermore, if states and their wars are the highest stage of the objective *Geist*, this, which includes politics, is then superseded by the three stages of the absolute *Geist* (religion, arts, and philosophy).

26. See illuminating examples of this link as regards France and the Duchy of Savoy in Tilly 1990, 26, 98. The class coalition and the political culture dominating in each country were decisive in shaping this process. Between the two extremes of

coercion-intensive (Russia, Brandenburg/Prussia) and capital-intensive (Genoa, Ragusa/Dubrovnik) ways of extracting resources, the combination of absolutist coercion and the promotion of manufacture and trade, according to Tilly, prevailed in the way chosen by major powers like France or England to feed their military budget. Modern war and war-preparedness were thus linked to the domestic social structure and political development. Class conflict as well as the origin of ideas like democracy and citizenship can hardly be understood if we do not make their link to foreign politics explicit.

27. No battle or siege of this cruel war, nor, centuries later, any of the battles between mass armies at Gettysburg, Verdun, or Stalingrad could have inspired literary accounts of war and battle like Goethe's *Kampagne in Frankreich* or Stendhal's description of Waterloo at the beginning of *La Chartreuse de Parme*.

28. Louis XIII of France and his ministers Richelieu and Mazarino are credited by Tilly 1990, 69, with having torn down more aristocratic castles than the fortresses they had Vauban and other military architects build at the frontiers. This symbolically telltale process towards a monopolized central force—towers in the interior down, towers on the border up—had begun three centuries earlier in the Italian *Comuni* and sea republics.

29. On the definition of these categories and their significance for the build-up of the polity cf. Schmitt 1927.

30. The exchange between the psychological (individual and/or group) level and the institutional level is recurrent in the fragile epistemology of a necessary, but ambiguous science like political psychology. Not even Freud was able to avoid its traps, and resorting to a questionable analogy he transferred patterns of individual psychology (the emotional ties which appease aggressiveness) to the states regarded as *Grossindividuen*. See the answer to the "why war?" question raised by Albert Einstein in Freud 1932.

31. I would add that, in order to be an institution, the regulation provided for by those rules must be "upheld by norms and by sanctions which are legitimized by these norms" (Eisenstadt 1968, 409).

32. This definition of institution will be helpful in the further discussion of this notion in chapter six, § 3, and chapter seven.

33. From sovereign statehood to the balance of power and later the "concert of nations," from *ius in bello* to the Red Cross.

34. Order is not identical with, but is supported by rules, in a relationship which is widely discussed by Bull 1977, chapter 3. The fundamental rule, a meta-institution or general principle rather than an institution itself, is *pacta sunt servanda*.

35. J.J. Moser was able to publish in three years (1777–1780) ten volumes of a compilation including "the most recent European law of nations," while Vattel's *Le droit de gens* (1758) refers to regulations in trade, navigation, diplomatic, and war-related activities; my considerations about international law are very largely indebted to Nardin 1983.

36. This term was introduced, according to Nardin, by Mably in his *Droit publique de l'Europe, fondé sur les traités* (1747), a title that highlights the shift from the "nat-

ural" to the positive foundation of international law. I use the Latin name in order to underline that commonality and formal equality under international law was only conceivable as far as the partners identified themselves as sharing the heritage—here I disagree with Nardin—of a superior language and juridical tradition (Roman law, and its developments in the Middle Ages). Self-images are powerful providers of justifications and motivations for the actors from/in the political as well as the legal system.

37. The identification of three main traditions of thought about states (plural)—Hobbesian/realist, Kantian/universalist, Grotian/internationalist—was first made by Martin Wight (1991, but this book is based on materials written decades before) and later developed by Hedley Bull (1977, 24–27, 322).

38. The best known example in the last decades was the U.S. government's ignoring of the ICJ ruling of 1984 concerning the mining of the Nicaraguan ports and coastline during the civil war between the Sandinista government and the contra rebels, which was a case of armed aggression against a state with which the United States was not at war.

39. This account relies on the reconstruction of the crisis made by Abram Chayes, the Harvard scholar who, as Legal Adviser to the State Department, was a member of Kennedy's "Executive Committee," see Chayes 1974.

40. How far circumstances can widely or totally change the role of international law can be perhaps visualized by some counterfactual comparisons. As for the personal and cultural background of the individual involved in the drama, things went the way they went in 1962 also because the man who had to make the decision (quarantine or blockade) was John F. Kennedy, not the forty-third U.S. president, and his closest adviser was a dove like his brother, Robert, not Henry Kissinger. The foreign policy of the country which had to make the first, decisive move had to reckon, unlike the Soviet Union, with a remarkable degree of public debate and democratic control, two elements which require policies to be justifiable in legal terms. However, they were not powerful enough to restrain pure *Machtpolitik*, preventing the Bay of Pigs adventure, or the Vietnam tragedy, or the subversion of Chile's democracy in 1973, or the ill-conceived and disastrous war on Iraq in 2003.

41. Effectiveness is the capacity of the system to achieve its goals; efficiency is the capacity to do so while minimizing costs and maximizing results. In both cases it is a matter of choosing the most adequate means: a problem of the technique of an action, as Max Weber dubbed it (Weber 1922, 65).

42. This image of (comprehensive) rationality is indebted to the redefinition of notions like reason, happiness, and materialism in the Critical Theory of the early Frankfurt School, cf. Horkheimer 1995.

43. If it does, as in Condorcet or Comte or some species of Marxism, we may speak of the "strong" model of Reason, and capitalize the letter R. Marx and Engels, although they certainly had a philosophy of history, were however suspicious of any "modern mythology, with its goddesses of justice, freedom, equality and brotherhood," as Marx wrote in a letter to F.A. Sorge on October 19, 1877, cf. Marx and Engels 1953, 364–65, translation mine.

44. In our time, the most comprehensive account of this relationship, whether or not its normative foundation of critical reason may be shared, remains Habermas 1984–87, particularly the first two chapters. Cf. also Toulmin 2001.

45. *Perpetual Peace* is the most explicit document of this stance, in particular the Appendix dedicated to the relationship of politics and morals.

46. This type is often labeled "cosmopolitical" in recent literature, but in the case of Kant this term cannot be used except in its very specific meaning related to *Weltbürgerrecht*. I am not happy with "universalism" either, but this is the current language, as established by the English School of International Relations. My account of the dichotomies of strategic/substantive reason and internationalist/universalist position has some points in common with Nardin's (practical/purposive attitude, cf. Nardin 1983); but the correspondence does not go that far, nor can I share his evaluation of the options ("practical" and "purposive" are irreconcilable with each other, the former is to be preferred). The discussion of cosmopolitanism will be continued in chapter seven.

47. Although Marx did not make this link a major theme in his research, he accurately described in the first book of *Capital* the unregulated and inhumane treatment of the expanding industrial working class at the time in which England's public debt increased from £22 million in 1697 to £238 million in 1783, after the Seven Years' War and the war with the United States of America (cf. Tilly 1990, 75). It can be presumed that only its roaring capitalist expansion allowed England to come to terms with this huge demand for financial resources. The French monarchy, which was unable to do so, instead had to convene the Estates General; the accumulated war debts thus played a relevant role in triggering the revolution.

48. This vision has indeed been taken seriously and implemented by millions of men and women in the history of the worker's movement (in the socialist as well as the communist wing) across the world, in peace and in war, such as the Spanish Civil War for example. It was bound to become the mockery we were used to from Stalin to Brezhnev, as the universalistic principle was identified with the interest of states, chiefly of the imperial power, the Soviet Union. This perhaps inevitable fate, up to the wars between socialist Leviathans in Southeast Asia in the 1970s, was facilitated by Marx and Marxism's incapacity to produce a critical theory of the state, and consequently, of interstate relations.

49. I may underevaluate the revolutionary and Napoleonic wars, but at their end in 1815 only two middle-sized states, the ancient Republics of Genoa and Venice, had disappeared. The two most relevant novelties (mass conscription and the ideologization of conflict, the latter already weakened in the campaigns of the *Empereur*) almost disappeared after Waterloo, and showed up again much later, in full strength as late as 1914. Perhaps what was more incisive, in the French Revolution and afterwards, was the fusion of nation and people in the idea of popular sovereignty, of which a *national* army was seen as the primary manifestation.

50. It is indeed difficult to provide empirical evidence for these theoretical assertions; even the cases of USA and Great Britain admittedly rely more on signs than actual proof. As for the new democracies in Western Europe after 1945, they indeed flourished under conditions of nuclear threat and ideological conflict, but they did so

in spite of these conditions. In Germany and Italy democracy became more deeply rooted and democratic change first occurred (center-left coalition in Italy 1963; Brandt's social-liberal coalition in West Germany 1969) as late as the sixties, as permitted by the first wave of detente between West and East.

51. As well as the people of the Americas (with the exception of the Civil War in the United States).

Chapter Three

The End of Political Modernity

Total and ideological war, genocide, and the possibility of nuclear omnicide as well as their counterpart, that is, the efforts made to institutionalize international society and its peaceful trends, have characterized the history of the last hundred years. Along with other social and political developments, such as the globalization of the economy and communication, they seem to amount to an end or a fading away of modernity in its structure and "legitimacy," as Hans Blumenberg (1983) terms it. What is seen in this chapter as being at stake is the rationality of the political order, a core element of modernity, since this was based on the claim that politics is a self-sustained sphere of human agency and can be rationally reshaped according to its own *ratio operandi*. This claim is put under severe stress by the evolution of the last and the present centuries, in which the global challenges to the survival of humanity or its civilization, which we have had to face since 1945, have made certain trends and threats, which had already come up in the thirty years before, escalate to new levels. In chapter four we will then examine another failure of the modern order, arising from the patterns with which modernity has addressed nature and human wealth rather than from politics, while challenging the modern idea of politics as well.

On the intellectual level, we are confronted with the task of defining the new situation and making it a topic for both philosophy and political theory. Neither of these disciplines can be said to have so far proven able to tackle the meaning of what is going on and to redefine their conceptions accordingly. It is as if our astonishment, as dwellers of the earth as well as scientists, about what has or might happen were too recent and too upsetting to allow for a swift redefinition of our conceptual tools. To give an example, anticipating a theme that will be developed in chapters five and six, social justice has long been and in some places is still the dominant topic in philosophy, occupying the place that should be given to the survival of humankind. Now, neither the sheer reaffirmation of the modern project of reason nor the postmodernist exit from a collective, if disenchanted approach to our future seems to provide an adequate answer. Two features of our epoch may have contributed to this embarrassment: the new aspect, which requires new insights and consists of the global challenges, is mixed with lingering mental and institutional patterns that originated in classical modernity, such as national sovereignty or unchecked confidence in technology and economic growth. Their resilience, which is a matter of fact, is wrongly taken as a justification for denying the need for major changes in theory. Furthermore, the "revolutionary" effect of novelties has been too often and too vainly proclaimed or predicted by philosophers and politicians, thus making caution highly advisable; this remains right so long as caution does not paralyze our nose for what is really unprecedented, as if "nukes" were just another discovery of gunpowder and global warming something as temporary and eliminable as the London smog of the 1950s.

The intention of these brief considerations is to underline the difficulties which surround any attempt to fulfill in a satisfactory way the first of those intellectual tasks: periodization. I am aware that my proposed scheme (classical modernity 1648–1914, its break-up from 1914–1945, the nuclear and global age from 1945 on) would need to be argued in a wider historical framework—and this could indeed be done. However, in a book which does not want to be a history book, it may be sufficient to suggest a periodization of the events that is consistent with the theoretical argument, which is focused on what "went wrong" starting in 1914. First came the shift towards total and ideological war, which led to genocide, however not for the first time in modernity, and established the possibility of nuclear omnicide (§ 1). Why, despite all its stability claims, nuclear deterrence was and remains a substantial disavowal of the rationality claims of modern politics is discussed in § 2, while in § 3 the weakening and fading away of this rationality is examined more in detail by looking at how some of its institutions (war, victory, neutrality, sovereignty, security, and democracy) are upset or thwarted. The roots of this evolution, primarily the link between security and technology, are laid bare in §§ 4–5 and seen as leading to the culmination of the security dilemma, also in connection with the overall crisis of modern strategic/instrumental as well as substantial rationality. What science and technology have meant for modern politics is further discussed in a historical perspective in § 6.

1. TOTAL WAR AND GENOCIDE

By *total* war I mean two aspects of the twentieth century's great wars:

1. The civil population is directly hit by war operations, either by way of "collateral damage" or because they are intended to be the very target, in order to force the enemy to step back or surrender. The big shift was the introduction of air warfare and the use of air power as an independent and ultimately decisive military instrument, anticipated in strategic doctrines (Giulio Douhet, Bill Mitchell) already in the 1920s. Together with the later-developed rocketry, air power stripped the territorial state of its "hard shell" and the civilian population of the protection which in the times of the "bracketed war" had been provided by the discrimination of non-combatants vs. combatants. Even if not actually bombed,[1] everybody in a country at war has the grounds to fear he or she will be: there is no "sanctuary" any more. The state is no longer able to filter and distribute fear and insecurity in such a way as to spare those not professionally committed to warfare. The globalization of menace in the age of ballistic nuclear missiles is the culmination of the process that dawned over Guernica in 1937; but the global scale now attained (everybody is under threat, not just in particularly exposed countries, but because of nuclear winter the world round, and would-be neutral countries are also included) creates a new quality. To a different degree of severity, this is true of the environmental global challenges as well.

2. Unlike the first point, the other aspect of total war was already fairly well developed in the First World War: the total mobilization of populace and social resources in a warring country. This meant conscription, general mobilization, and the build-up of huge armies, followed by huge war cemeteries and huge numbers of crippled veterans. On the other side, the economy and social life was first put under control and later regulated by the state in a way that was unknown to the purely private capitalism of the nineteenth century. Thus, in Europe warfare accelerated the shift from the liberal and *laissez-faire* individualism of the bourgeoisie, portrayed by the classical French and German novelists (from Balzac to Thomas Mann's *Buddenbrooks*), to a bureaucratic and unregulated monopoly capitalism, which was largely intertwined with a high level of public bureaucracy and military elites.[2] This was a quick and effective way to extend the modern labor process along with the resulting "iron cage" (Weber 1905, 181) to the whole of society, thus fostering modernization in all its aspects: disruption of

traditionalism and localism (the emancipation of women in the twentieth century is inextricably linked to total warfare) as well as bureaucratic, often authoritarian social control. The "mass society," which attracted the main attention of social philosophers like Ortega y Gasset and Emil Lederer between the two world wars (and later that of David Riesman), was first forged by the organizational upheavals that were required to build up and to feed the mass armies of the Great War.

The ideologization of war, admittedly an ambivalent notion, pushed its total character to an unprecedented and so far unthinkable degree, although the two features are conceptually distinct from each other. Nationalism alone was never able to raise the emotional mobilization and the group psychological glue later provided for by such eschatological pictures of a future world order based on metahistorical values such as the rebirth of the Roman Empire envisioned by Mussolini or Hitler's *tausendjähriges Reich*.[3] In the first half of the twentieth century war thus ceased to be an acceptable and quasi value-free means of settling disputes, a feature that in the framework of *ius publicum europaeum* was previously symbolized by the institutionalized "declaration of war." B was no longer, in A's view, *justus hostis*, a rightful adversary, but rather an aggressor or oppressor not to be simply fought against, but condemned and punished. Let us remember that, looking at the Treaty of Versailles, particularly at Art. 227, which established the German Emperor's responsibility for the "supreme offence against international morality and sanctity of treaties," Carl Schmitt lamented the further mortal blow brought to *ius publicum europaeum*, specifically to its non-discriminatory concept of war (Schmitt 1950, 259–80). Let us put aside the question of Germany and Austria-Hungary's *Kriegsschuld*: Schmitt failed in any case to see the methodological point that the war with victors and vanquished, but no culprits, could not survive under the new circumstances (deadly, non-"bracketed" duels between great powers; mass destruction weapons; total commitment of a country to the war effort). Last but not least, the hope that countries governed by democracy or based on the rule of law would not initiate wars was widespread among liberals and democratic socialists: war could no longer be regarded as a value-free instrument of politics, peace became a value not just in morals, but politics as well, and waging war should be regarded as an international crime. This change of mind eventually became generally accepted and institutionally endorsed in the UN Charter as well as in post-1945 national constitutions (first of all in the countries which had triggered the Second World War: Japan, Italy, and both German states). Faced with the unprecedented levels of destruction of contemporary war, the European tradition of universalistic reason experienced its own powerlessness in

Ethiopia in 1935, Spain in 1936, the whole of Europe in 1939 and achieved at the same time a final, not merely moral triumph over the realist practice of states:[4] war became an illegitimate means of regulating international affairs, except for strictly defined (Art. 51 of the UN Charter) self-defense or interstate police operations under chapter vii of the Charter. As a backlash to wars fostered by totalitarian and aggressive ideologies, and beginning, at the latest,[5] with the Spanish Civil War, the stakes of war happened to be dictatorship and terror vs. freedom, like in Spain or Nazi-occupied Europe, or foreign rule and exploitation vs. (however flawed) self-rule, like in the Cuban and Algerian Revolution, in Vietnam and Afghanistan under Soviet occupation. These are alternatives of fundamental significance for both the moral and political field and cannot be reduced to "ideology" as mere coverage for traditional power struggles, although they can be misused in this sense. What we have called the *ideologization of war* thus turns out to be a complex feature, which includes several and conflicting factors, all inevitably bound to put war under an axiological assessment: it can mean world views and political ideologies aiming at civilizational dominance rather than territorial gains as motives for waging war, or an unprecedented level of destructiveness multiplied by ideological mobilization, but also an improved consciousness of how political life should be reasonably organized in a peaceful way.

Now, a war like World War II, which was at the same time total and ideological, did not simply expose the civil population to the effects of military operations like strategic bombing and the guerrilla warfare of resistance movements in almost all of the Nazi-occupied countries along with the cruel antipartisan warfare. Beyond the logic of war operations (or against it, as those historians argue who stress the costs that the German war effort had to pay to the "final solution" machinery), a new instrument of pure destruction, specifically devised against civilians, was set up in the Third Reich. Primarily directed at the Jewish people, it also took, by the hundreds of thousands, the lives of political opponents, Slavs, gypsies, homosexuals—not to speak of the Russian prisoners of war. It is not necessary here to enter the debate on whether the "final solution" or Shoah or Holocaust was originally foreseen by the Nazi war strategy, or came about as a reaction to the failure to obtain a final victory against the Soviet Union: military operations and genocides were interrelated aspects of Nazi politics.

In this regard, Auschwitz represents a cleavage in modernity which has distinctive grounds, but no lesser significance than Hiroshima: the new dimension, including effective and planned genocide as well as omnicide as an unintended, but accepted consequence, was introduced into civilization at the same time and in close connection with each other.[6] Both were made possible by scientific and technological development; however, at Auschwitz the primary factor, at the latest after the Wannsee conference of Nazi officials (January 1942), was the genocidal intention, which only needed to find suitable technology (the Zyklon-B gas and the crematoriums) and organizational techniques (first the SS-Einsatzgruppen, later the network of camps, industrial locations like the Buna-Werke at Auschwitz-Birkenau and transportation facilities). The potential omnicidal effects of nuclear weapons were not at all intended as such: ironically enough, the A-bomb was built to prevent Hitler's genocidal politics from acquiring that immensely powerful instrument which, however, was at that time seen as a bigger bomb rather than as an epoch-making invention. Furthermore, whatever one may think of the decision to build the H-bomb and its leading proponent, Edward Teller, among the decision-makers the aim was not to exterminate all the Russians or all the Communists, but to prevent them from getting there first. However, what is nuclear deterrence indeed if not the idea of threatening the enemy with genocide and humankind, including one's own people, with omnicide, in order to avoid it? Genocide of the enemy's population, although a very real potentiality, is in any case meant not as an end in itself, but merely in conditional mode: "we do not intend it, but it is up to you not to force us to resort to it in order to stop or to punish you." Here it is not the political or ideological intention, but the *nature of the war means* which, beside or even against any intention, may bring about genocidal or omnicidal results. Later in this book I will resume this theme in its relevance for both philosophical (the relationship between means and ends; science and technology) and political theory (the role of institutions).

The Third Reich was defeated and destroyed, but genocide or attempts in this direction are far from disappearing, as seen in the nineties in the former Yugoslavia, a country in Europe, and the African region of the Great Lakes. Only the industrial scale of genocide has so far been dismissed, and less sophisticated methods, similar to those used by the Turks in 1914–15 in the Armenian genocide, have been reinstated. The peaceful integration among former enemies in Europe, the main positive novelty in world history after 1945, has been tainted by the inability of the new Europe to prevent or to stifle war and genocide, the promise on which European integration was built ("no more war" and "no more Auschwitz"), in the Balkans. The recent enlargement of the European Union to the former Communist countries of Eastern Europe, while decisively improving security and democratic cooperation among the peoples of this continent, makes it look like a war-free "realm of relative reason." However, even regardless of what can happen on this continent with nuclear armaments, all this cannot cancel the fact that genocide has briefly reappeared on the European continent in the early 1990s and is touted as a legitimate intention elsewhere.[7]

But genocide, or at least the lack of effective restraint in this regard, also lies in the backyard of our history, at the exact time that our continent was governed by Schmitt's "relative reason." If we look back at the course of modernity with unbiased, non-Eurocentric eyes, we see not one, but two genocides being carried out by Europeans and their descendants at the same time in which the "bracketing of war" and the *ius inter gentes* began to take shape on our continent:

1. the purposive slaying, directly or through disease, alcohol, or famine, of millions of native Americans, not to speak of other colonial peoples;
2. the enslavement of about eleven million black Africans between 1550 and 1880 and their deportation, with a high death toll, to the Americas; not to speak of those deported eastwards by Arab dealers.

Beyond their extra-European "amity lines" (Schmitt 1950, 92–98), the European states did not only fight against each other in a less well-educated manner. They chiefly regarded the life and death of entire groups of the human kind as pure functions of their longing for free land and a cheap labor force. Killing or enslaving them for this functional purpose was facilitated by the various versions of racist ideology which the European mind developed in reaction to its first encounter with the cultural and anthropological otherness of "Redskins" and "Negroes" (cf. Todorov 1999). With the killing of tens of millions of Europeans by the hand of other Europeans in the twentieth century, it is as if a cynical dialectics had brought back into a much more highly developed Europe that same tool of power and pain that Europeans had first used against other peoples and races.

A terminological uncertainty at which I have already hinted can now be clarified before ending this section. I have been using "genocide" with regard to attempts made by states or other actors (slave dealers between Africa and the Americas, interest coalitions around the American frontier) to destroy either well-defined (national, ethnic, religious) groups or humankind. I have dubbed "genocide" both the effect of an open intent (Jews, Armenians, to a certain extent American Indians) and the accepted consequence of actions having other purposes (the exploitation of slave labor, the deterrence of the nuclear enemy). These actions or attitudes are different from a legal or historical point of view; but for moral and political philosophy they have in common the acceptance of *"asymmetrical" extermination as a means* to pursue collective goals. Asymmetrical is the kind of extermination wherein the victim does not threaten the perpetrator, nor would actually be able to inflict the same treatment upon him; this makes this kind of extermination (which I have referred to under the more familiar term "genocide") different from exterminations taking place during wars, or at least during wars in which the *ius in bello* principle of the proportionality of means is respected. It can be objected that nuclear war between nearly equal powers (or more strictly speaking: nuclear war between actors in possession of second strike capability) cannot be regarded under this definition as extermination or genocide. This is true in narrow terms, but on the other hand nuclear war is genocidal or rather omnicidal *a fortiori*, since it includes the possibility of suicide for one's own people. In nuclear war strategy, the human life of an extremely large group of individuals (inhabitants of whole continents or the human race altogether) is rated even less than in genocide proper, in which the extermination of another group, foolish and ruthless though it may be, is perpetrated for the advantage of one's own group. As victory is impossible, large-scale nuclear war lacks even this rationale, thus turning to be even more "nihilistic" (cf. chapter five, § 4) towards human life, or rather the life of humanity, than the narrowly defined genocide.

2. NUCLEAR WEAPONS AND NUCLEAR DETERRENCE

Inevitably, the nuclear weapons issue has been already touched upon several times in this book. I shall now deal in a more orderly manner with those of its countless aspects which are relevant for my philosophical argument. We have to find some tentative answers to questions like: Where do nuclear weapons come from? Second, to what extent did they matter in post-WWII history and, third, how reliable is the reassuring insistence on their political-only nature as tools of deterrence?

To begin with, nuclear weapons are a legitimate child of the twentieth century as well as of modernity in general, which is not equal to saying that they unveil the true nature of modernity, science and Enlightenment. There would be no nuclear technology without nuclear (as well as atomic and particle) physics. There would be no physics whatsoever without the specific modern attitude towards scientific knowledge, especially the interplay of theory and experiment. This applies specifically to the developments that paved the way for nuclear technology, from Rutherford's studies on the atomic nucleus (around 1911) to Otto Hahn and Fritz Strassman's discovery made in 1938 of how the uranium atom splits under neutron bombardment. These works had nothing to do with technological applicability, were motivated by mere scientific interests and conducted at a low cost within universities or research institutions with no need for special funds provided by the military or industry. The physics that made the A-bomb possible was thus a pure product of (mainly European) scientific rationality. Its other presuppositions were the technological and organizational ability of modern states, based on either a market or command economy,[8] to provide the necessary resources and push the process forwards towards military application, once the state had engaged in total war. Upon these presuppositions, in other times it would have simply taken the political decision of the state to put the process into motion and to produce the new means of its power. This was not so with the Manhattan project: however interesting its history from the vantage point of political decision-making (not just to build, but to drop the A-bomb) may be, it is not this element that furnished the *causa efficiens* for the bomb to really be built. The best officers of the U.S. Army Corps of Engineers plus the best R&D departments in American and British corporations would not have been able to do it; it took something more than organizations like these run by the common impersonal codes of power and money, or command and efficiency. It was first necessary to gather on a largely voluntary and personal basis the best American and refugee (later also British) physicists, transforming them into engineers working on the application of their own scientific discoveries. To attain this goal, a different impulse was needed and was in effect provided by a further feature of the twentieth-century cleavage: ideological war, whose complex meaning I have sketched before. The decisive motivation for the scientists gathering at Los Alamos was the will to build the A-bomb before the Third Reich was able to do the same.[9] Furthermore: the motivation to anticipate any Nazi bomb was so overwhelming, particularly among the European refugees, many of whom were Jews, that moral concerns with the utterly destructive effects of the new weapon were superseded by the political and moral intention of defeating Hitler.[10] This role of the *ideological motivation* is stressed by the fact that, once Nazi Germany was defeated, a group of physicists of the Manhattan Project, based chiefly in Chicago and led by Leo Szilard, asked the U.S. government not to use the A-bomb against cities (Bundy 1988, 95; Weisskopf 1991); earlier, two researchers had left Los Alamos because of the end of the Nazi threat (Weisskopf 1991). Thus we have the origins of the nuclear age from the political and scientific point of view. On the ethical and cultural level it can be briefly mentioned that the framework of an old moral dilemma, the "tragic choice" situation, was thus filled as early as in the summer of 1945 with new contents of an intensity never seen before. In a definition of this notion different from Calabresi and Bobbitt's (1978), it is the choice we ought to make by comparing not good and evil, but two different evils (say, ending the war in the Pacific by erasing two Japanese cities or landing and fighting with enormous, Okinawa-like casualties on both sides), and choosing what we deem to be the lesser evil under conditions of cognitive uncertainty, which makes us somehow feel guilt on two counts, first because what we choose is still an evil, even if perhaps a lesser one, and then because we feel that we may be wrong in assessing what the lesser evil is.[11]

Let us now turn to the second question. While the *political history of nuclear weapons* cannot be examined in this book except for single aspects or episodes, two general remarks need to be made at this point. First, the balance of terror and nuclear deterrence, which I will be dealing with as the basic structure of the nuclear age, was quite obviously not contemporary with the first appearance of nuclear weapons on the political stage (1945). This developed later in the fifties on a new technical basis: the thermonuclear bomb (first exploded by the United States in 1952, and by the Soviet Union in 1953) and the intercontinental ballistic missile (ICBM, first deployed by the United States in 1959). What was later called the MAD strategy was first devised in the years 1952–54 in the framework of Eisenhower's *New Look* (cf. Bundy 1988, 246–60): Ike's "massive retaliation" was certainly different from McNamara's MAD, but also established a major precondition to it, as it implied the accumulation of warheads and delivery systems. But the nature of the nuclear bomb as an absolute and overwhelming novelty had been recognized by Bernard Brodie as early as in 1946 (Brodie 1946): one more reason for considering 1945 a cleavage. On the other hand, the further step (because of its unprecedented destructiveness, the bomb can only be a "political" weapon for deterrence, or a last resort, if deterrence fails) was, as an effective policy, made much later: in the late fifties or early sixties. Still in the Korean war and shortly before the French debacle at Dien Bien Phu (1954), the military use of nuclear bombs on the North Korean–Chinese and Vietnamese forces was generally considered possible, though it was then rejected for traditional reasons of opportunity; among them, the small amount of available warheads.[12] Besides, in the forties and fifties, the exhaustion of all major countries (especially the Soviet Union) in WWII as well as the recent memory of it, a deterrent for peoples and governments, played the overwhelming role in preventing war.

The second remark is aimed at qualifying that attribute of stability in a more contrafactual than historical way. Nearly forty years after Winston Churchill's speech of March 1955 on the consequences of the hydrogen bomb, had his assumption that peace might be "the sturdy child of terror" proved right or wrong? On the one hand, right: beyond the general fact that the Third World War did not take place, a number of international crises of such severity can be recalled that in prenuclear times would probably have led to a military conflict between the great powers. Even if we exclude from consideration episodes of the fifties like those already listed and others (Hungary and Suez, 1956; Quemoy and Matsu, 1958), let us first mention the Cuban missile crisis of 1962, the sole and the most relevant case in which nuclear weapons were determinant both at the origin (Soviet installations on Cuba) and the solution (restraint in both sides' behavior for fear of a nuclear war) of the crisis. We should also remember the Berlin crisis of 1961, as the city was divided by the wall built by the government of the German Democratic Republic, and the "alert crisis" during the Yom Kippur War of 1973, which directly involved the U.S. nuclear forces. In all these cases the restraint imposed by nuclear deterrence can be credited as having averted both nuclear and conventional war. In another case conventional fighting erupted between nuclear powers at the Chinese-Soviet border, but remained far from becoming nuclear.

Assessing the *importance of nuclear weapons and nuclear deterrence in recent history* is not an easy task. It is not only that their interpretation will remain an open question as long as all of the great powers' archives have not been made fully accessible. It is also difficult to separate the role played by nuclear deterrence from other factors and considerations of a political, military (conventional), economic, and moral nature. These factors may have included the nuclear weapon not as a military component in the power relationship, but rather as a symbolic message in diplomatic communication; for example in the "alert crisis" of 1973 it was used as a symbol of America's determination to keep Soviet troops out of the Middle East.[13] In other cases nuclear weapons may have not been used because of the assumed lack of political support, both internationally and in the domestic public sphere (Bundy 1988, 537), rather than because of the fear of retaliation, which was unlikely to occur. Following a suggestion made by Bundy, I would speak of a self-deterrent effect of nuclear weapons: even in situations (Korea, Vietnam) where they were tactically quite usable and advantageous from a military point of view, their use was deterred because their use was considered morally unacceptable by the very power who would have used them.[14]

As for the third question, concerning the stability and instability of the nuclear situation, interpretations are made problematic by two possible flaws, the first of which derives from the risk of mistaking declaratory policies for the real state of affairs in nuclear power relationships.[15] Furthermore, what we to some extent know and can discuss are American, as well as British and French, policies and doctrines; a comparative history of the Western and Soviet nuclear strategies is still to be written. However, we already know enough about the Soviet leaders hardly sharing Western-born conceptions, like "limited nuclear war," or (what is more) the meaning that Western leaders attached to certain steps they were taking in both nuclear doctrine and policy.[16] These doctrines and policies should be reassessed in the light of the impact they had or failed to have on the opposing party, taking into account not only the classic catalog of misperceptions examined by Jervis (1976), but also the fact that in the Cold War the distortions of language due to the ideological contrast did very much add to the semantic differences that in any case result from different cultures; hence also the delays in correctly interpreting the new policies of the adversary.

It is a widespread persuasion among scholars of the nuclear age that nuclear stability needs a *second strike capability*, which deters first strikes or surprise attacks.[17] This capability was and still is based on a "triad" of delivery devices: intercontinental Ballistic Missiles (ICBM, launched from the ground), bombers (in the United States, the B-52) and submarine-launched ballistic missiles (SLBMs). SLBMs were and still are typical second strike arms, because they can be regarded as untouchable by a surprise "decapitation" strike and are inaccurate enough as to better fit counter-city than counterforce targeting; this is thought to deter the enemy from starting a war which would in any case, by retaliation, inflict upon his population and economy insufferable pain.[18] On the American side, second strike capability can be said to have been firmly assured since 1959 and 1960, as the first ICBM and the first Polaris boat became operational.

Yet not even the availability of weapon systems allowing for second strike capability guarantee stability. There were two sources of instability in the Cold War: first, political change (as well as turnover of political personnel) in the domestic sphere of a nuclear power, or in countries under its influence, for example in the Third World, to the extent this change was brought to bear on nuclear policies.[19] The other, internal source was the strategic doctrine that, like declaratory policy and/or effective technical change, prevailed at different times. Some doctrines may have favored a higher grade of stability, some others a lesser one; but all of them contained their own germs of instability. Let us look at the case of Eisenhower's threat of "massive retaliation" in the event of Soviet attack (even in a proxy war). We do not know if it actually deterred the Soviets from initiating those actions, but what soon became evident was that it left no option beyond the "general nuclear war" or "loss of face" alternative, thus reducing the room for credible deterrence and flexible action. These flaws of any massive use (or threat of use) of nuclear weapons repeatedly aroused criticism from people advocating a role for those weapons in a more articulated view of international conflict, according so to say to a back-to-Clausewitz view of nuclear war as "real" (*wirklicher*), not as "absolute," that is unlimited war. This idea of a "limited nuclear war," first surfacing in the debates of the late fifties, then again in the seventies, was pursued, in terms of both viability and damage limitation, as an alternative to all-out nuclear war; in a general sense, McNamara's "flexible response" falls under the same headline. But two conflicting aspects made this idea highly ambivalent: on the one side it was developed as a better deterrent due to its greater credibility (greater than the credibility of a massive threat). While on this side it can be credited with stabilizing effects, on the other side the same feature, that is, viability, which should make it a better deterrent, may sound as if it is actually encouraging the resort to war, a thought involving destabilizing effects indeed.

Let us now briefly reconsider the stability-instability issue in the light of more abstract concepts, which have been credited with being influential on operative doctrines, even if the amount of their influence cannot be unanimously determined: MAD, "escalation dominance" (Hermann Kahn) and "a threat leaving something to chance" (Thomas Schelling). Mutual Assured Destruction through counter-city targeting supported by second strike capability became a central feature in the second phase (starting in 1964) of McNamara's tenure at the Pentagon. Whatever the history of McNamara's formulation may have been (cf. Freedman 2003, particularly chapter 16, 232–42), MAD can in a more general form be regarded as the core structure of international politics in the nuclear age, in which the fear of one's own intolerable damages in the end has kept the nuclear powers from exploding nuclear weapons in any given situation, for example, in a surprise attack, or to compel others to do something; or to compel them to stop doing something, for example, to cause the Soviet Union and China to stop helping North Vietnam in the Indochina wars. After the end of the first phase of the nuclear age (marked by the end of ideological and imperial bipolarism as the basis for nuclear bipolarism), we can observe that, as far as nuclear deterrence has worked, it has not done so thanks to particular policies and doctrines, and at times even in spite of them. Rather what has caused deterrence to work has been:

a. the fear of one's own destruction, which, whatever the strategies adopted, everybody had reason to believe would be, even if not exactly mutual and assured, in any case enormous and unrewarding;
b. the unprecedented uncertainties about the way in which a nuclear conflict would develop, primarily about the chances for its limitation, albeit boldly advocated by some doctrines;
c. to a far lesser, but not irrelevant degree, the incompatibility of (even a "victorious") nuclear war with one's own moral and cultural standards.

Stability through the sheer fear of self-destruction seems thus to be the attitude which best fits arms which have outgrown their nature as instruments of (limited) destruction on behalf of political aims.[20] But it is the yet unavoidable involvement of those arms in the power play called politics, which can but reintroduce elements of instability in nuclear deterrence. However

wide the consent may be on the "absolute weapon" being a device of pure deterrence, the temptation to resort to it as a "relative" means (i.e. related to political goals of victory or blackmail)[21] must from time to time come up again
in the political and military mind: is not, after all, the possibility of its actual
use, as well as the technical care for its efficiency and credibility, inscribed in
the very concept of deterrence? In this regard, the political use of nuclear arms
for pure deterrence cannot be seen as separate from their military use for fighting and winning a war: these two extremes are distinct, like the poles of a
"magnetic" relationship, rather than separate. We need not primarily refer to
war-fighting doctrines or policies in order to demonstrate the powerful presence of instability in nuclear history. First of all, we should rather look at the
presence in deterrence doctrines, though he was by no means a theorist of pure
deterrence, of Hermann Kahn's idea of dominance which has to be built up and
maintained on every rung of the escalation ladder as a guarantee of deterrence
stability and an instrument for asserting one's own political goal. This inevitable search for superiority generates instability, not only inasmuch as it
promotes the arms race; what is more destabilizing are the provocations which
can be seen by the adversary in any step taken to establish or re-establish dominance in the escalation, as well as the countermeasures he may resort to.[22]

Finally, uncertainty and instability are not presumable side effects, but
rather consciously built-in factors in the most sophisticated of the doctrines
which have been devised to improve deterrence stability: the case made by
Thomas Schelling for a "threat that leaves something to chance," a chapter in
his seminal *The Strategy of Conflict*, the book that is a key to nuclear policy
in the same way as *The Prince* was a key to the politics of early modernity
(Schelling 1960). Dismissing the idea of a plan or policy capable of covering
every contingency, Schelling argues that the most effective way to deter an
enemy from doing something we dislike is to keep him guessing about our response, and tell him that whether and how we may respond is not completely
a matter for us to decide upon. The extreme form of this attitude is to go to
the "brink of war."[23]

While assessing the stability and instability of nuclear deterrence during
the past decades, a reasonable compromise should be brokered between competing views. We should acknowledge that deterrence worked, even if it has
been exceedingly credited with a peacekeeping capability, which may have
hinged upon other, non-nuclear factors as well. It did not work by mere
serendipity, but rather through a learning and rationalization process, which
some statesmen, strategists, and military men completed earlier and with a
firmer sense of restraint than others. Yet in nuclear deterrence the structural
component of (objective, not consciously introduced) uncertainty and
brinkmanship should likewise be recognized: the failure of bipolar deterrence

has always been a built-in possibility. To infer from the bloodless end of the Cold War the confidence that nuclear deterrence will not cease working, thus relieving us from the task of thinking about alternatives, is tantamount to the behavior of a person who, after surviving a single game of Russian roulette, decides that "it isn't dangerous after all" (Schelling 1960, 190). On the contrary "the proposition that nuclear weapons can be retained in perpetuity and never used—accidentally or by decision—defies credibility" (The Canberra Commission 1997).

This critical and hardly reassuring view on deterrence[24] is the presupposition we need in the next section in order to grasp how deeply nuclear arms and deterrence have upset the political rationality of modern times.

3. THE COLLAPSE OF THE RATIONAL INSTITUTIONS

Let us look now analytically at how in the nuclear era the institutions which made the modern political world an ordered and self-sustaining structure have been made void or radically transformed. Those institutions were created to meet the challenge of a world upset by technological, economic and cultural novelties such as geographic discoveries, the introduction of firearms, the improvement of navigational techniques, the reshuffling of production, and trading in capitalist sense. The fundamental new institution was the modern state, the Leviathan, but I shall deal with the categories in which its novelty found expression, and they are war, victory, neutrality, sovereignty, security, fear, and, on the side, democracy. I shall not explicitly deal with peace, because the change occurring in it is sufficiently described by the indications pertaining to the other categories. Some of the modifications I am now going to touch upon are merely potential, and would only become actual if a nuclear war occurred. Others are illustrated in a definitive form, although the presence of nuclear weapons and its unprecedented effects coexist with the permanence of previous "modern" structures and styles of action. On the one hand, I look at the loss of function of those categories in the international system. This is on the other hand complemented by the loss of meaning which affects those institutions, whose integrative force for the members of political society is decreasing or vanishing. This can happen as soon as, under nuclear circumstances, the meaninglessness of fighting and dying for the country, or keeping loyalty to the state as provider of security to present and future citizens becomes evident.

Let us now take a closer look at the destiny of those institutions.

a. War as a resource for self-assertion and power redistribution has been outplayed by nuclear weapons: it is no longer a resource to which the political system can freely and unlimitedly resort when other means of self-regulation appear to have failed. Originally intended to be but a major step in such modern and scientific undertaking as strategic bombardment,[25] the nuclear bomb has turned out to be a device which would forcibly make true Clausewitz's "absolute war" (cf. Bull 1977, 50), in a way which would indeed be destructive for *all* parties involved, whatever the combatants' intentions. The international system has thus been deprived of what used to be, for example, in the balance of power system, an important instrument of self-adjustment. This subtraction of an important element from the international system would of course be no loss in (substantive) rationality, if the resort to (nuclear) war had been completely banned from the system, and more peaceful and reasonable instruments had taken its place; but there is no assurance of things having reached this point of no return. The nuclear nightmare has only fostered a substantial move forwards in the process which began in the trenches of the Great War and has been stripping war of attributes like beneficial, progressive, and necessary (cf. Mueller 1991, 65). But a new firm order of war and peace is still to come; in the best case we are going through a long transition. Even more time is requested in cases of the (pretty uncertain) assumption that war, primarily nuclear war, first becomes "rationally unthinkable" for the logic of politics, and then "subrationally unthinkable" in the logic of cultural evolution.[26] In any case, to assume a cultural and even anthropological trend heading towards the disappearance of war, as happened with duels and slavery (Kaysen 1991, 82–83), is no substitute for the task, which is specific to political theory, of finding out the principles, the institutions, and the timing under which the process capable of banning nuclear war can be pursued and consolidated. After all, one could argue, humanity as a whole has coexisted with slavery for thousands of years and has not been hurt, except morally; the only ones who were really hurt were the enslaved minority. Yet it is not as easy to assume that it can coexist with the threat of nuclear war, which can at anytime come true and blow out the vast majority of human beings. As for dueling, unfortunately no "domestic analogy" can be assumed between individuals and states, as, unlike individuals, so far states have been unable to learn how to settle their disputes in a court of law, instead of by violence.

On a more abstract level, nuclear weaponry has shifted the balance which existed between compelling and deterring, the two aspects of power in general. This gives the opportunity to clarify how this fundamental notion is used throughout this book. If we define *power* as the relationship among actors in which an actor or a group of actors induces other actors to act in a way in which they would not otherwise act,[27] this relationship can further be seen either as deterrence, in which the other actors are kept from doing what they have not been doing, or of compulsion, in which they are induced to do what they have not been doing or to stop doing what they have been doing. Military power and war used to exert both forms of power,[28] establishing a *continuum* between them. Quite differently, nuclear power cannot be used for compulsion, at least not between nuclear countries, as it is only apt to exert deterrence; the political power game must resort to forms of compulsion (economic competition, imperial influence and proxy wars, terrorism, as happened in the Cold War) other than nuclear or at most to conventional war between the major nuclear powers: this may have created Mueller's distorted view of nuclear arms as being "essentially irrelevant," whereas they are the defining terms of the present world situation.

Two further changes in war need to be briefly examined: the relationship of war and politics from the venture point of *military technology* and the problem of war termination. Nuclear warfare is regarded as the culmination of the contemporary trend towards "automatic warfare" (Basil Liddell Hart, quoted by Freedman 2003, 20), which excludes fighting based on military and civic virtues, and turns to pure destruction. Now, as far as automatic warfare contains the chance of bringing military power under the strict surveillance of the political leadership, it can, in a democratic country, enhance control over military shortsightedness and institutional irresponsibility. Complaints sound like antimodern, romantic nostalgia for the good old warfare of Azincourt or Austerlitz and deserve the same reply that Hegel gave to the "chivalrous" criticism of firearms as allowing the most cowardly fighter to slay the most gallant one.[29] The other side of the coin is the process which, while concentrating the decision-making process in only a few hands, takes substantial decisions out of the political sphere of competence and makes them, for example, launching on warning, the outcome of electronic procedures, leaving little room for reflection, communication, or bargaining.[30] As for *war termination*, this used to be the result of a complex relationship between military and political (diplomatic, domestic) elements. Though the decision, made by Churchill and Roosevelt at the Casablanca Conference in 1943, to put an end to World War II only upon the unconditional surrender of Germany, Italy, and Japan, looks like a purely military termination, it was indeed based on historical and political considerations. On the contrary, the problem of war termination is practically non-existent after a wide-ranging nuclear war which would have disrupted entire continents, their society, and political structure. In its own presumable dynamic, nuclear war seems to be an open-ended event, which can hardly be brought back under political definition and goal-setting.

b. War termination used to mean *victory* for one of the players, but "a nuclear war cannot be won and must never be fought." This statement made by the otherwise "muscular" President Ronald Reagan (for the first time in 1982, during the arms race between the Soviet SS-20 and the American Pershing II and Cruise missiles in Europe, see Bundy 1988, 583) was the open acknowledgment of a latent, but powerful truth. If nuclear war occurs, the "winner," that is, the side which inflicts upon the enemy more damage than it receives, will itself suffer to such an extent that it will not be able to credibly claim "victory"; perhaps there will not even be a central authority capable of raising this pretension. War outcome has thus ceased to be the basis for reshaping the international distribution of power and prestige, in a way that used to last for decades, as it did after 1815 or 1870 or 1918. Since 1945 change in power relationships, for example, the power gained by the countries of the European Union, chiefly by Germany and by Japan, and lost by Russia, has not occurred by means of war. But it should as well be remarked that this change has been confined to the economic sphere and has not reached, as would have happened fifty or one hundred years ago, the military one: between the two spheres there is now a divide which did not exist in earlier times, and neither of them can gain full control of the political one. In other words, since war has been suspended as an actual regulator and replaced by nuclear deterrence, there seems to be no direct and complete transformation of new economic and cultural power into the political sphere. Nuclear war and victory in war can no longer redress the "balance of power," a notion whose temporary revival has therefore little to do with balance of power proper, and should not be mentioned except between quotation marks. As for collective security, that is, the regime which since the end of World War I was expected to replace the obsolete balance of power in regulating the international system (cf. Claude 1962), it can now rely on deterrence alone, as far as great powers are concerned, and not on war.

The fact that military victory has vanished in the nuclear world does not mean that nuclear arms play no role at all in a political contest which may still end with the "victory" of one side, as happened at the end of the Cold War. Its comprehensive history is still to be written, but in the posture of witnesses we can say that, if nuclear weapons played no role in terms of a direct competition (i.e. the Soviet Union did not step back in fear of the West's nuclear superiority), they did so in two other regards: they kept the competition substantially (proxy wars aside) bloodless and chiefly based on economic, political (leadership), and cultural (legitimacy) factors. Even more importantly, they shaped the competition as a process between two blocks. These were not based on nuclear power alone, but it was the capacity of the two block leaders to build up a strategic nuclear strength that gave those blocks a worldwide dimension and imperial structure. At the very least, strategic superiority over the other nuclear and non-nuclear countries sealed the Soviet-US bipolarism and made it more effective and, for a long time, more stable than would have been possible through economic power or ideological pretensions alone. This structure was shaken by factors other than nuclear, although the political weight of the Euromissiles race is not to be underestimated. But was the crisis of the Soviet Union not due to its being overburdened with an imperial role towards which its ideology pushed and its nuclear capacity led, or perhaps forced it?

c. One more institution tied to war has vanished under the effect of nuclear weapons: *neutrality*. This has technical and political reasons. Technically, the side effects of a large nuclear war such as nuclear winter, would heavily, perhaps lethally, affect most or all of the countries in the globe, or at least those near to the core theater of a nuclear exchange, with no regard to their diplomatic stand: an important aspect of the *ius publicum europaeum* has thus been blown out by the nuclear wind. Politically, it is in the nature of strategic nuclear power, in spite of the will of would-be neutral countries, to make the build-up of territorially vast alliances a necessity, in order to provide a "nuclear umbrella" even for those countries which may in other issues not completely align with their leading powers in a bi- or *n*-polar system (where *n* is not greater than 4 or 5, since numbers matter in this issue!).[31]

d. Along with neutrality the very *sovereignty* of the state is affected. Its crisis and transformation shall be discussed in chapter seven, but the nuclear component of this process needs to be outlined here. Sovereignty as *summa potestas* of the individual state A over the fundamental decisions affecting its destiny fades away when decisions, which simply because of their technical range may concern the survival of this state and its citizens, are made by other states (in fact, by their military and political elite), to which A has assumed either a neutral posture or the attitude of a "sovereign" ally. How side effects of nuclear explosions can affect neutral states has been already mentioned; in the case of alliances, like NATO and *a fortiori* the Warsaw Pact, it is a constant in history for the leading country to have an overwhelming weight in making important decisions.[32] This weight has also been enhanced by technical aspects of decision-making in nuclear war or deterrence. Against this view one can put the stress on the free and autonomous act of even entering an alliance on the side of a country which then puts its destiny in the hands of the most powerful partner, or of an allied command, as has always happened since the Trojan or Peloponnesian War. But what was at stake in traditional alliances was merely the strength and prestige of the state, rarely its survival as a state, and never the survival of its citizenry and civilization. Besides, opting out of an alliance was possible at any time, although sometimes heavy costs had to be paid—as happened with Italy after the armistice with the Allies in September 1943. Opting out of an alliance during a nuclear war would have at the very best the effect of being hit by nuclear winter–like effects alone, instead of being directly targeted. The geographic basis of modern sovereignty, a "hard shell" surrounded by a "buffer zone," breaks apart under the threat of nuclear attack by ICBMs and SLBMs or even delivery systems with a regional range, which do not respect Carl Schmitt's geopolitical separation of land and sea (cf. Schmitt 1950) and hugely multiplies the effect already created by air power and strategic bombing.

This is the most significant attack on the sovereignty of the modern state, but it is far from being the only one: I shall come back to these matters in chapter four and seven. The sovereign state has indeed been the legal framework in which modern rationality has elaborated its two fundamental regulations, regarding security and the domestic distribution of power, the latter's other name being democracy. This has been affected by nuclear weapons not just because these tend to engender an undemocratic "guardianship," as Robert Dahl (1985) argued, but rather in the radical sense that through the erosion of the role of politics they bring about, and the enormity of their effects in the event of war they make the very promise of democracy untrue, that is, the citizens' destiny being entrusted to the citizens' community. We will resume this discussion in chapter seven, so let us now turn to examining how the nuclear revolution has affected the category of security.

e. Three Hobbesian postulates are no longer true: the state can no longer guarantee *security* as its basic performance in favor of the citizens' industry, because the industry as well as the life of nearly all citizens would be lost if it comes to actual fighting between "gladiators" who are armed not with sword and shield, but with ICBMs and SLBMs: the losses that the "gladiator" has to suffer in order to ensure the citizens' security are no longer justified if the last is thereby not only not enhanced, but rather put in danger. It has also become untrue that security is as asymmetrically distributed as to allow the most powerful actors to believe themselves basically preserved from destruction: even the *force de frappe* of a minor country can inflict upon them insufferable damages. Thirdly, the emergency exit that Hobbes kept open for the worst case, namely when a state is no longer able to provide security because it has been severely beaten or dissolved in a war, does not exist anymore. If there is no victor in nuclear war, but devastated countries in which complex organizations like the state can scarcely exist, the chance of the citizens saving themselves by shifting their allegiance to the new victorious sovereign as Hobbes suggested, vanishes as well.

With these postulates losing their ground, the very meaning of the social covenant becomes doubtful: if the state is not able to provide the security on whose behalf it has been (in the contractarian view) created, why should citizens still recognize it? It is not so much a question of revoking the transfer of all potentials of violence into the state's hands; what can be revoked lies a layer underneath the state's monopoly of violence, and can be indicated as the interest of the people in being active partners of the covenant, that is, participating citizens. If the people perceive the practical lack of meaning of the state as supreme protector, they need not show their feeling by revolting, but simply by abstaining from voting and other political and community activities: "exit" behavior, whose motivation is empirically not easy to locate, since it usually has multiple sources.[33]

The nuclear situation has thus undermined the validity of the traditional idea of security and ushered in a new one, still in the context of sovereign nuclear states. Traditionally, the distribution of the security resource is *disjunctive*: I am secure insofar as you are not, that is, security by superiority. When regarded from the outside, collective security among the partners of an alliance still belongs to this type. But if the costs of enforcing security by fighting back are too high for all the actors, then *common security* is needed, that is, reciprocal behavior in which everybody, while keeping his deterrent, tries to accommodate the security needs not just of his partners, but of his potential enemies as well, thus creating a regime of confidence and detente and enhancing everybody's chances of surviving and thriving.[34] A corollary to the idea of common security is the necessity to shift from a one-sided, mostly military security to *comprehensive* or *integrated security*, which combines diplomatic, territorial, and military factors with a minimum of economic (water, food, and energy supply) security for every actor, even hostile. This type of security was not unknown, at least among great powers, in times before the nuclear age, and has been partly implemented between East (or rather Russia) and West after the Cold War; on the other hand, it can be argued that without more distributive justice, international order must remain unstable. But under nuclear conditions it appears to be a necessity for survival rather than a better option among others, although this can be misused in the direction of admitting only nuclear countries to the benefits of, say, economic security; a circumstance that creates one more motive for leaving the countries that happen to be as poor as nuclear-free out of the door or pushing them to become nuclear.

4. THE ROOTS OF THE PROCESS

If it is true that fundamental categories of modern politics are made void or forced to change their meaning, the roots of this process are likely to lie in deep underlying layers of the recent evolution of civilization and polity. The scenario for this evolution is shaped on the one hand by the shift towards total war among sovereign states, and on the other hand by the scientific discovery of how to detect and take into war service the ultimate forces of matter and energy. There is, however, a third feature in this scenario: the tendency, which was very pronounced during the Cold War, to defer political security to *technology*, or to replace the social and cultural communication endeavors aimed at establishing a comprehensive security by more and more refined technical devices, which are believed to be an impassable dam against the adversary's will and weapons. This trend can be referred, on the one hand, to the general cultural trend towards *deferring to automated procedures and technical objects* the management of social situations which, because of their complexity and repetitiousness, would otherwise overburden our capacity for decision-making. Instead of dealing with the complexity of the problems raised by the adversary as well as the world situation, thus making both our mentality and our interests flexible, we may prefer to put order in this situation by affirming our independence and superiority in a way which we deem guaranteed by things that we have, while the others do not. This possibility to simplify the situation is even more welcome if we already have a trend towards simplifying the image of the adversary and reducing real problems to products of his malicious attitude: in this sense nuclear security perfectly dovetails with ideological war and cultural Manichaeism.[35]

This tendency is somehow coupled with a *reification* of security, which has been largely or completely entrusted not to political skills, but rather to things, that is, the possession of an amount of certain things which always has to be higher than the amount in the hands of the adversary. Since a reified idea of security is not interested in any verifiable knowledge of both the adversary's real attitude and the overall situation, the tendency to regard the "worst case" as the most likely is inherent to it, and the adversary is assumed to gain possession of more and/or better things than we do. Instability and the arms race were therefore deeply rooted in nuclear-ideological bipolarism. A further aspect of reified security[36] is the idea that, if we have plenty of smart things and keep pace with the others, steadily inventing smarter things, we may achieve a comfortable, perhaps *absolute security*. That this result is unfeasible and attempts made to achieve it may unleash a pre-emptive strike from the adversary is not the sole reason that this idea is a gloomy security utopia. In the Cold War it came together with the unwillingness to come to terms with uncertainty and contingency under nuclear conditions, and in the several stages of the arms race led to the effort to escape this new, terrifying level of uncertainty by false means, that is, by reproducing and enhancing exactly that situation, nuclear weapons in the anarchic states system, which is the root of terror.

Automation, reification, and absolutization of security under the auspices of unprecedented technological leaps: this means the theoretical focus needs to be shifted to technology itself. My approach to the technology problem is however different from the widespread thesis that technology has an autonomous life, away from human control, as I believe that technology does not get out of control as a result of its own might, but rather appears to become autonomous as long as its developers, the human beings, are not able to invent a new cultural frame of reference and to establish *new institutions*, within which they can come to terms with the social and political challenges raised by technological leaps forward. This problem will be discussed at length in chapter six, § 3. But in the nuclear age the link between technology and institutional change has specific features and is no mere illustration of a general rule. While during the Industrial Revolution the delay in devising and implementing a new institutional frame ended up internationally in a zero-sum game and domestically in asymmetrical burden-sharing,[37] extending the *time* needed for the institutional regulation of the nuclear issue may bring about two effects. First, the danger of nuclear weapons being used in a nuclear war, including a regional war, grows as time goes by, because the longer the pendulum swings between deterrence and possible war, the higher the probability of a swing going as far as the actual use of those weapons. Furthermore, this probability is increased by the shift from a relatively stable situation like bipolarism to a more unstable one, like horizontal proliferation, which is what has happened after the Cold War with the increased appetite of regional powers for nuclear weapons as a tool of supremacy (Iraq before 1991) and/or symbol of prestige (Iran at the time of writing) or reassurance against regime change (North Korea and again Iran). This is peculiar to a type of technology like nuclear weapons, which produces social consequences in a discrete mode, that is, cleavages like Hiroshima and Nagasaki or points of no return like all-out war, rather than by continuous accumulation like, say, the steam engine. Secondly, the costs to be paid if nuclear technology is used for fighting a war are not as asymmetrically distributed, as it used to be in conventional wars, since they inflict unbearable burdens on all humankind, in a game whose outcome is negative for every player. In this case, the need for institutional reshaping is stronger and more urgent than with other technologies affecting the existing order. This is true not just of nuclear arms, since it can be referred to all technological developments generating global challenges, as confirmed by the (relatively fortunate) episode of the ozone layer-depleting CFCs (cf. chapter four). It can perhaps be suggested, but not argued in this book, that the fact that patterns of order are more often than previously affected by technological change is a specific sign of modernity, in which this change is unprecedented in volume and swiftness; while the fact that order can be upset and deprived of legitimacy by the effects of a scientific-technical event signals a cleavage in modernity or its end altogether. With these theses I should not be mistaken for saying, in general philosophy of history mode, that all upsetting of social order stems from technological change; it is but one source of change, even in modernity (technological change played nearly as small a part in the French Revolution as it did in the fall of the Roman Empire).

If my assumptions about the link between technology and security are right, they may shed some light on the questions of the "causes of war" in the nuclear age and ask for a change in the traditional doctrine towards a "fourth image" of the relationship between war and politics, pointing at the new constellation of war, politics, and science/technology beyond Waltz's "third image," which looked at the international states system (cf. Waltz 1954). Let us proceed in a counterfactual way: nuclear conflict has not taken place so far, but if it had, none of the traditional causes of war would have been a satisfactory answer to the question of how it was possible. Economic expansionism or, put bluntly, imperialism, geopolitical factors, tensions between rich and poor on a world scale, the military-industrial or bureaucratic-military complex, the influence of the military on the ruling class (including military rule or dictatorship), diplomatic manipulation, nationalist propaganda and ethnic hatred, domestic turmoil in a superpower or bloc: all these traditional causes of war, some of which were at work during the Cold War, are too rational an explanation for such an unprecedented oddity like the balance of terror, let alone nuclear war. Causes like these seem to arise from contrasting interests, which can be solved with costs for everybody, but also more benefits for the winner, and to motivate armed conflicts which are mostly as long and as cruel as needed to settle the dispute and give the winner the gains he or she was seeking; sometimes cruelty can go as far as required by the traditions of the winning tribe or country. In World War II already (to some extent also in its precedents, the Spanish Civil War and the Japanese invasion of China) cruelty and destructiveness went much further than was required by the necessity of war or the conflicting parties' cultures: the Japanese tradition did not make the Nanking massacre necessary, nor did the German culture prefigure the Jewish Shoah, even if we regard genocide as a possibility contained in the modern mind.[38] In what can be assumed to be the evolution of a nuclear war, an unprecedented amount of destructiveness would be unleashed, an amount incomparably huger than in World War II, yet without hatred, unlike in Auschwitz, and rather with a kind of chilling indifference for the lives of millions of men and women whether friends or foes. Insofar as we can make any sense of this potential event, it seems to be the result of an obsessive search for security, in particular for a kind of security which is entrusted to "things" such as bombs and regulated by automated devices.

Our "fourth image" needs to be further explained. The search for security is constitutive of every polity; but if this search becomes obsessive, aiming at absolute security, there need to be some special reasons. Some authors, like Kaufmann 1973, have put it down to the dissolution of the organic ties among individuals alongside the development of a mass society in the industrial era. The isolated situation of the individual in the abstract net of "society" (Tönnies' *Gesellschaft*), as opposed to his or her being embedded in the "community" (*Gemeinschaft*), is regarded as the source of his or her search for security, both in the domestic sphere of social policy and in international affairs. This hypothesis tackles some aspects of the recent evolutions, but it seems to be too general (external and domestic security can hardly be referred to the same mechanisms of individual psychology) as to be satisfactory in our case. The obsessive traits of the search for security during the Cold War appear to have had two closely connected roots: first the strong ideological Manichaeism which was mirrored in the several "images of the enemy" on both sides of the Iron Curtain (cf. Weart 1988), then the simple fact that the new weapons had made the stakes immensely higher in material as well as psychological terms, because the people had to be reassured against total annihilation rather than mere defeat or invasion, and the new technological wonders seemed to do the job. The traditional security system was thus upset in the first place by a factor, like the advance of natural science in the twentieth century, which was widely, as we have seen, an independent variable, rather than a product of political developments, and happened to bear poisoned fruits exactly at the same time as the poison of extremist and chiliastic ideologies.

In this constellation, along with the lack of a new political arrangement capable of coming to terms with it, security through increasingly automated deterrence devices seemed to be the only viable way of managing a situation in which neither side accepted either to be politically and technologically overwhelmed, or to go to war, the traditional solution that a supreme power struggle used to require. In this context the role of ideological Manichaeism should not be overrated: it was of critical importance in the genesis of the nuclear situation, helped to kill early attempts to reach a global agreement,[39] and it contributed to blocking or delaying steps towards detente and disarmament. But it cannot be said to have been the sole pillar of the balance of terror: once the nuclear arms race was initiated, it developed a logic of its own, that is, the nearly cyclical motion between political and military use of those arms we have described above. Technical innovations, like ballistic and cruise missiles, nuclear submarines, anti-ballistic missiles (ABMs), missiles with multiple warheads (MIRVs), and satellites, have had at times both a destabilizing and a reinforcing effect on the deterrence regime which went beyond the intentions heeded in the political decision behind them. Even in this case, the political impact of a new technology is still, at least in part, an independent

variable: it cannot be wholly calculated in advance and made fully instrumental to a political design. On the contrary, it puts constraints on the latter.
These patterns of interaction can subsist even if the ideological breeding of
the opposition between two countries or blocs subsides and is replaced by
other motivations, like imperial competition, or ethnic or religious struggle.
They can be altered in a peaceful way only if a deep change in the contenders'
political cultures as well as in their institutional bonds occurs: this is what
happened in the eighties with *perestroika* in the Soviet Union and the INF,
START, and CFE treaties between the East and West. But on the whole we
must say, using popular terms, that since our kind, like an apprentice wizard
of the twentieth century, fell victim to the temptation to base its security exclusively on deterrence and the latter on technology, it got involved in a game
whose rules were and still are out of its hands, as the present deterioration of
the non-proliferation regime demonstrates.

5. THE NUCLEAR SECURITY DILEMMA AND ITS CONSEQUENCES ON RATIONALITY

Even those who may not share the scheme of interaction between technology
and politics I have just outlined will recognize that the outcome of this interaction is clear: the nuclear revolution has upset the *institutional regulation of fear
and security* which in the contractarian account used to be the very core of the
modern state. Fear has been seen as originating from sources like a disrupting
social change, cultural revolutions (the world losing its centrality in modern as-
tronomy, or the society becoming too complex) or an acute scarcity of basic resources. Politics, which managed to build up a centralized, rationally organized
state, put restraints on fear, inasmuch as it tried to contain and to regulate its expression in social unrest, or to prevent its upsurge by eliminating or stabilizing
its sources. Except under pathological conditions like tyranny, the state would
neutralize all fear and only generate *timor legis*, the equal fear of the law, whose
internal structure is different from pre-political fear.

The novelties upsetting this modern regulation are that my fear is now inspired by my country's as well as my enemy's (nuclear) arms; that, because
of its new, overwhelming military might, the state (or rather the system of
states) has acquired the capability to generate more fear, the fear of reciprocal annihilation, and a more generalized, inescapable fear that used to exist in
the state of nature, that is, in civil disorder before the establishment of the
modern state or even in international anarchy before the establishment of the
still anarchical society of states that was conceptualized by Hedley Bull. If indeed the institution which is supposed to provide self-preservation and secu-

rity for any of its members fails to do so, if the net sum of security vs. insecurity is negative, then the very sense and justification of this institution is undermined. This happens, in other words, whenever Leviathan, the fearless sovereign, becomes as frightening as his opposite, the Behemoth of civil disorder. It is first at this stage of evolution that the *security dilemma*, a notion I discussed in chapter two with regard to Hobbes' theory of the modern state, comes to its culmination.

The outcome of the security dilemma in the nuclear age has been regarded in this chapter as a factor which can undermine the state's legitimacy and weaken or disrupt the bonds of social contract. But this topic calls for differentiation. At a first glance, empirical evidence does not provide much support for it. After half a century of nuclear history, states cannot be said to be manifestly deprived of legitimacy and allegiance because of their involvement in nuclear policies. Nonetheless, behavioral patterns or mental phenomena signaling "exit" tendencies from the social and political life under nuclear threat have been investigated from different scientific vantage points.[40] What is more, sometimes evidence does not emerge because, in the absence of a new theory addressing the new phenomena, nobody has raised the right questions. If the political scientists who examined the state's legitimacy crisis solely in terms of its domestic or social roots had broken the intellectual division of labor and integrated what happens among nations into their methodology, more empirical research would be available and shed light on our topic; the same holds true for the phenomena widespread in the West of disaffection towards public institutions and political life, one of whose roots may lie in the perception of their inability to come to terms with global threats. But a further methodological point has to be made: legitimacy, as a category of political philosophy, is not identical with consensus, which can be empirically tested at any time. This is said on the one hand to justify the direction of my research, that is, the unfolding of a theoretical argument about the state's own legitimacy being potentially undermined by its mentioned inability, even if so far this potentiality has not become actual; nor will it do so in predictable times, because with the end of the Cold War the nuclear nightmare seems to have vanished from people's minds. On the other hand, I am trying to keep away from a simplistic argument of the kind "what is theoretically predictable will sooner or later come into existence and drive peoples and states." With regard to fear and its institutional regulation, we must be aware of the factors which exclude direct political consequences: first, political behavior is motivated by reactions to events which are relevant in the people's everyday life rather than by insights into background structures or threats, which may be perceived from time to time, but also tend to be repressed or become latent. Then, the arguments presented by the theory of narcissism about the recent weakening of passions and the neutralization of fear in the motivational structure of social behavior must be taken into account.[41] Thirdly, the real situation has changed: even if we keep the possibility of nuclear weapons again playing a major role among world powers open, as in my view we have to, this is unlikely to happen in the next five to ten years, the lapse of time with which political science on a empirical-analytical basis is usually concerned.

This reveals a difference in the salience of nuclear threat for philosophy and political analysis. However, it is a difference that the nature of the threat may be about to reduce. Political science has been able to keep its distance from philosophical issues raised about our destiny as humankind as long as it had to reckon with a condition defined by *moderate challenges*, as I argued at the end of § 2: no nuclear winter, no diversion of the Gulf Stream was in sight. The very magnitude (physical as well as moral or cultural) of global threats shortens the gap between political problems and philosophical perspectives: from time to time those threats may emerge or disappear from the political stage, but this stage itself is permanently surrounded and overshadowed by their presence and the explicit or tacit concerns they raise. Political actors or theorists can no longer resort to the traditional commonplace of dismissing such concerns as "philosophical," that is, irrelevant for policy-making. The differences that we have briefly discussed should advise an unbiased, fallibilistic approach to the question of the political weight of global challenges: the possibility of them (as they emerge in philosophical consideration) becoming actual political issues should be regarded as a true possibility, that is, something that may or may not turn into actuality, rather than something that will necessarily either occur or disappear. Nor should we pretend to transfer into political theory the absolute and dramatic shape they are seen in when we look at their significance for civilization. A prudent intertwinement of philosophical and political views is nevertheless advisable if we want to make our political culture fit to meet the unprecedented requirements of the nuclear age. For the passage from moderate to global challenges that the political order is going through seems to be the cleavage dividing modern political rationality from what may come after it.

The thesis of global challenges representing the end of modern political rationality can be questioned as it may appear to be too general. What is poised to collapse, one could say, is the substantive rationality which in history saw the nevertheless conflict-laden march of Reason towards the attainment of higher goals like freedom or justice; in short, what is in danger is the picture of progress designed by several philosophers of history. It would go too far to include in this collapse strategic rationality,[42] which on the contrary can see its own triumph in the enhancement of technical efficiency in warfare and elsewhere. It is not by sticking to the deeply controversial notion of rationality, our critic could argue, that we can understand the significance of global challenges, nor find out what needs to be done.

A first reply could be given in these terms: instrumental or strategic rationality is expected to find the most effective and less costly means to an end, that is, in our case, the defense of one's own country or civilization. If the means chosen brings about the destruction of the country, the course of action was strategically (Weber would say: technically) irrational.[43] A second, less polemical reply should tackle the issue of *substantive* vs. *strategic rationality* in its full complexity, but in order to discuss this thoroughly I would have to once again take up the main chapters in the interpretation of modernity, from the Enlightenment to Weber, from Marx to Habermas. As this cannot be done here, I shall limit myself to a somewhat reductive formulation of theses, or rather hypotheses, as this is hardly a field where certainty, either rationalistic or catastrophist, can be seriously proclaimed.

The first point to be made is that both kinds of rationality were joint in the original "project of modernity," their separation being a later result of processes which allow for various explanations: the inner dynamics of formal rationality, along with the impossibility of fully subjugating reality to rationality (Weber), or the unfolding of the hidden antagonistic core of the capital's *ratio* (Marx), or the ambiguous scope of the rationalization dynamics of our "life world" (Habermas 1984–87).[44] For sure, the separation between strategic and comprehensive, or systemic and life-world rationality (Habermas) has occurred, and cannot be reversed by reminding us of their original unity. But this can shed light on some presuppositions that strategic rationality tacitly requires in order to be able to work and to justify itself. These presuppositions, which imply an original *unity of the rationalization endeavors*, are:

- intentionality: actions are intended to reach certain goals, thus reckoning on "the future" as an essential dimension;
- trust: although strategic action is based on "using other human beings for our own goals," not even this is possible without a certain amount of prudent reliance of one on each other, that is, on the existence of other human beings who can be convinced, manipulated, influenced, and who in a relatively stable frame of action can act as we require or expect.

Both presuppositions, the existence of the future and the existence of humankind as the dimension in which the goals of our rational strategies can be attained, are no longer guaranteed in the nuclear era, thus undermining the general frame of reference in which strategic rationality can work and keep its promises. This is why we cannot say that all claims made by substantive rationality are dismissed by global challenges, while the strategic one remains unharmed. Or can they be totally separated from each other? We can indeed imagine someone acting with perfect strategic rationality, but subjugating it to the will of Beelzebub, or of a Freudian Thanatos totally deprived of Eros' company, and thus purposely pointing at the elimination of humankind. The organizers of the Nazi network of extermination camps as well as of the GU-LAG Archipelago are cases in point. But their efficiency was limited to the single issue of extermination and repression, and the political systems they were part of were defeated either on the battlefield (Nazi Germany) or in imperial competition (the Soviet Union) by countries whose higher efficiency was surrounded by a *reasonable* sense of humanity. As for political actors such as the superpowers in the Cold War, on the one hand their nuclear arms race can be regarded as a typical example of strategic rationality leading to the destruction of humankind if we look at what the objective outcome of the arms race might have brought about. Nevertheless we cannot forget that, with regard to the actors' intentions, the strategic rationality of developing more and more destructive weapons was not intended as an end in itself, but rather was embedded in the allegedly "reasonable" intention of strengthening the deterrence regime, thus avoiding nuclear war. What becomes visible in this entanglement is that strategic rationality (pursued by actors in the arms race) can have unforeseen evil effects (the destruction of civilization) in spite of the actor's claim of pursuing reasonable aims (self-defense, pure deterrence as the precondition of nuclear peace). This is an argument in favor of an attitude joining strategic rationality with responsibility, an ethical and political category which global challenges urge us to adopt, as we will see in chapter five.

6. POLITICS AND TECHNOLOGY

Let us take another look at the evolution that we have gone through in this chapter. In modern times, peace and war among nations were regulated in a relatively stable and reasonable way, thus avoiding major destruction. Beginning in 1914, the very institutions of modern political rationality evaporated or turned into the opposite, and in the nuclear era a new regulation which can make claims of rationality credible is still far from emerging. The following question must be raised: how could relative rationality turn into full destructiveness?

I have so far suggested that the explanation lies in the huge leaps forward made by technology, particularly the technology of warfare and genocide, as well as in the lack of political institutions capable of averting its most destructive use. This explanation works on the basis of the comprehensive and multilayered notion of rationality unfolded in chapter two, § 6. With its help

I have tried to make sense of Carl Schmitt's "realm of relative reason" formula for the early modern regulation of peace and war. But I have not properly answered the question of why, along with that regulation, such a rational construct also collapsed. I do not intend to resort to general philosophical explanations of why something went wrong in modernity: either I do not deem them true (as in the case of value-nihilism[45] or the "death of God" being the source of all evil in modern history), or they have such a general (as in the case of Horkheimer's "eclipse of reason") or such a particular, monodisciplinary character (as in the case of Marx's critique of capitalist industrialization) as to be able to account for what occurs in the specific realm of politics. Explanatory elements can be drawn from all or almost all of these accounts, but whatever direction of research we choose, it will be unsatisfactory if it fails to explain the immense increase in destructiveness brought about by technology. However important, the other factors that have shaken modern political rationality, like worldwide industrialization, expansion of the state machinery, and the struggle of ideologies, have not had the same dramatic impact on the life of nations and humankind.

Again leaving aside for later discussion the philosophical assumptions about technology (cf. chapter six), we should perhaps focus on two structural aspects of the relationship between technology and politics. First, under the specific modern constellation of technology, politics, and economy, scientific and technological innovation progresses at a different, much higher rate than cultural change, including the change of political institutions. For example, since the introduction of the mechanical loom and steam engine, cultural (social, political) modernization has been experiencing a chronic delay with respect to the modernization of science, technology, and industry. Until the social and political institutions of the countries in which a capitalist economy was established were restructured, establishing democracy and social legislation, the working class had to pay the costs for the time gap between the introduction of technological innovations in industry and the cultural change that took note of its effects and finally introduced public health care. The fact that international institutions have not been created or reformed in such a timely and effective way as to meet the challenge of total, industrialized war and, later, nuclear weapons can also be understood with reference to that basic *pattern of delay*. The costs to be paid for that have indeed been huge and, with the last leaps forward in destructiveness made by technology, they could become catastrophic: once this threshold has been reached, the structural delay itself becomes a challenge to present and future politics, in the sense that it should be substantially reduced by speeding up cultural change, by a timely assessment of the consequences that may derive from technological innovations, and by the policy decisions to invest in technology development A rather than B or C.[46] *Technology is no longer morally or politically value-free*, and politics is confronted with the new task of governing its development by regulating and complementing the market (in the case of greenhouse emissions) and, in the case of nuclear arms, the states system, which as such are both far from being fully able to provide humankind with reasonable and timely solutions.

The second structural moment in the relationship between technology and politics that I want to highlight is an empirical rule, a pattern of political behavior: no technical means which gives hope of preserving (H-bomb) or increasing (firearms, even though chivalry despised them; A-bomb) one's own power is rejected or given up by a political actor. This *topos* of realism explains why a piece of technology, once its potential with regard to military or economic power has become manifest, acquires something of a normative character and *must* be adopted. It is a peculiar feature of the nuclear age that this rule is now weakened and exceptions become possible whenever all or nearly all of the interested actors develop a cooperative (post-anarchical) framework in which they renounce the new technology under institutionalized (international treaties or regimes) conditions of trust. However, it is remarkable that the exceptions concern a type of technology, like nuclear weaponry, which is not altogether set aside, but rather restricted to a few powers; furthermore, this happens only *after*, long after it has caused bitter experience, as predictions about its impact were not sufficient to move political decision-makers to early acts of restraint.[47]

The two modalities of the technology-politics relationship I have just mentioned do not answer the question "where does technological change come from?"; nor do I intend to embark upon a general discussion of this topic, also because I think that no monocausal explanation can be given, since technological change may stem from political decision or economic impulse in the same way as it can emerge from "uninterested" scientific research, as an opportunity that is within easy reach, as soon as we decide to transform basic scientific knowledge into technical devices. For example, what people like Lord Rutherford, Paul Dirac, Enrico Fermi, Otto Hahn, and Lise Meitner did was rather driven by the pure "joy of insight," a successful climax in the unfolding of "theoretical curiosity"[48] whose emergence in the Renaissance marked the cleavage between the Middle Ages, which restrained *curiositas*, and the modern age. Science is obviously intertwined with economy, society, and politics, but it is sociological reductionism to believe that each of its phases or steps must be accounted for as a mere reflex of class or power interests. Sometimes it can work as an independent variable, which raises unexpected opportunities or challenges for economy or politics. But this topic can be better clarified after introducing the specific features that result from man-made climate change in chapter four, § 4. On the whole, we can regard nuclear physics, nuclear technology, and nuclear weapons as a telling and mature, if threatening product of the whole course of modernity. The novelty is that one of the main actors, the Leviathan, no longer seems able to match the unexpected challenge resulting from a historical development in which it played a major role as a fighter and collector of resources for fighting with other Leviathans. Modern political rationality began to collapse as soon as it was confronted with an utterly destructive technology whose possibility was implied in the cultural and political structure of modernity, even though historically it emerged from scientific events that did not derive from political decisions. Signs of collapse also come to light in another story, in which the engine of troubles is located in the economic approach to nature, in the modern gargantuan appetite for fossil energy rather than in the political relationships among human beings and their communities. After examining the specifically *political* dialectics of modern Enlightenment we now turn our attention to a different chapter in it, in which the acquisition of greater wealth for humankind and particularly its most successful members threatens to doom the living conditions of future generations of men and women.

NOTES

1. Like this writer, and many European children during the Second World War.

2. As portrayed in the paintings of Otto Dix and Georg Grosz, or in Heinrich Mann's novel *Der Untertan* (later filmed by Wolfgang Staudte as *The Kaiser's Lackey*) and morally analyzed in the aphorisms of Horkheimer's early work *Dämmerung* (1978).

3. Communist ideology, also with regard to Nolte's (1987) debatable thesis of a "European civil war," would require a more complex assessment, which cannot be developed here.

4. This triumph of the opposition to war in the last hundred years has hardly been without a down side. In the thirties the search for peace went so far as to appease dictatorial and aggressive states, with disastrous results for world peace. This can perhaps be seen as the counterintentional, but hardly avoidable outcome of approaching war and peace in a purely universalistic manner, losing any interaction with both the internationalist and realist view, cf. chapter seven.

5. The American Civil War (in which the slaves' emancipation eventually became a major issue) and the First World War (in which the "democratic spirit against German authoritarianism" formula played a role) can be regarded as ushering in this change.

6. I am aware of the fact that calling the effects of nuclear bombing a genocide does not fit in with the definition of genocide as framed in Art. 2 of the 1948

Convention on the Prevention and Punishment of the Crime of Genocide: "acts committed with intent to destroy, in whole or part, a national, ethnic, racial or religious group as such" (Henkin 1980, 324). This definition is clearly modeled after the acts of genocide committed by the Nazis; even the term "genocide" was first used in this context by Lenkin in 1944. The restrictive clause "with intent" makes it legally difficult to recognize as genocide many episodes which clearly amount to it. Likewise, the lack of limitation to an identifiable political or social group among the victims of genocide means the killing of millions perpetrated by the Stalinist or Khmer Rouge regimes cannot be identified as genocide.

On the other hand, the A-bombs dropped by the United States over Japan in 1945, which ushered in the potentially omnicidal nuclear era, were not employed with genocidal intentions, since it was not the widespread popular hatred for the "Japs" that motivated the decision.

7. Think of the promise to erase Israel from the geographical map repeatedly made in 2006 by the president of the Islamic Republic of Iran, a country that may go nuclear in a few years' time.

8. Perhaps the first success could only be achieved in the most modernized capitalist country, and under the push of a personality like Robert Oppenheimer; but even the ability of the Soviet Union to catch up in very few years, despite the drawbacks caused in nuclear R&D by the Stalinist system, was astonishing evidence of what a modern (Soviet socialism was also a modernizing force) state apparatus is able to accomplish, even if we discount the advantage provided by atomic espionage.

9. Whether the Nazis' effort had chances and how long it lasted is now subject to a vast amount of literature centered around the life and times of Werner Heisenberg. In the years (mid-sixties) I spent at Frankfurt in Jürgen Habermas' research group on a fellowship of the Alexander von Humboldt Foundation, which Heisenberg chaired with a humanistic attitude, his role during the war was widely unknown or suppressed.

10. According to Weisskopf's recollection (Weisskopf 1991) and in private conversation with this author (Newton, Mass., spring 1993) it was first upon Niels Bohr's arrival at Los Alamos in the autumn of 1944 that the moral concern was voiced. Bohr spread around the hope that the new devastating weapon would soon induce its actual (USA) or potential (Soviet Union) owner to join an international control agreement, thus putting an end to war for good.

11. The decision to drop the first atomic bombs was a typical case of tragic choice, while there is not much serious historical evidence for a Machiavellian plot aiming at an unnecessary use of the weapon in order to win the first round in the power game between the United States and the Soviet Union. See, however, in this sense Alperovitz 1965; *contra* Brodie 1973, 56, and Bundy 1988, chapter 2, particularly note 78, as well as Fussell 1988, 1–28, though this last author makes no reference to Alperovitz's thesis. A later balanced account is given by Herken 1988, also with regard to the complex question of a mere demonstrative use of the A-bomb.

12. This is the explanation given by MacIsaac 1986. President Eisenhower seems to have perceived the moral impossibility of handling the nuclear bomb as if it were just another weapon, as he rejected the idea of using "those awful things against Asians for the second time in less than ten years" (Bundy 1988, 270).

13. The difficulty of weighing the (causal) relevance of a single factor or event in historical explanation (*Erklären*) is a classical problem in the epistemology of the social sciences, cf. Max Weber's discussion of the significance of the Marathon battle in the development of Western civilization in Weber 1906 and Weber 1922, 11. On the "alert crisis" cf. Lucarelli 1994.

14. As long as the question is not further clarified by the historians of the Korean war, we can assume interplay between moral or political restraint and military opportunity as the reason for the bomb not being dropped there. For the Vietnam war see Bundy 1988, 536.

15. For example Mutual Assured Destruction existed, as a real effect of weapons accumulation, a long time before it was proclaimed by Robert McNamara in the mid-sixties, cf. Freedman 2003, 234.

16. Cf. Freedman 2003 on the Soviet attitude rejecting the escalation doctrine and assuming that nuclear war would be all-out war from the very beginning (chapter 17, 243–57), or finding little sense in the idea of selective targeting (chapter 25, 355–77).

17. This is the background against which all Strategic Defense Initiatives are bound to sound destabilizing among strategic players, beyond probably being unfeasible, as it has turned out to be for the past twenty years. Deterrence stability vanishes if I know that my retaliatory capability is checked by the enemy's anti-missile shield, and that he has nothing to fear; once his window of vulnerability is closed, he might even launch a first strike, thus forcing me to take pre-emptive steps, in any case making arms control pointless. What a shield may mean in front of regional or terrorist threats cannot be discussed here.

18. This "unacceptable level of destruction" was quantified in the sixties by Robert McNamara, who indicated in the destruction of 20% of the population and 50% of the industrial capacity of the Soviet Union the immediate effect of an American strike, Freedman 2003, 234.

19. Examples are Khrushchev in Cuba 1962 and, later, the Soviet missile build-up after his fall in 1964; more generally, the coupling of arms control talks with the international behavior of the partners. Cf. Powaski 1987, 180–81 on the effects of the Soviet invasion of Afghanistan. On Khrushchev's adventurous Cuba policy, now see Fursenko and Naftali 2006.

20. Fear of even one nuclear explosion and mistrust about the possibility of containing a nuclear war once an exchange had been initiated were, according to Lebow and Gross Stein (1994), the effective forces of deterrence and restraint during the Cold War, rather than deterrence doctrines that for some time had exacerbated the existing tension.

21. I am not using Brodie's labeling of the A-bomb as the "absolute weapon" exclusively in the meaning he gave to it (a weapon whose efficacy is not relative to the number of units available to a country). I am attaching to "absolute" a second meaning: nuclear arms are, because of their enormous and suicidal effect of destruction, not relative to usual political goals (a means to a goal), but rather an end in themselves, i.e. something *finibus solutum*.

22. The Soviet nuclear build-up in the seventies, which led to the new Euromissiles arms race, can be interpreted as a reaction to their own strategic inferiority, as perceived during the Cuba crisis, cf. MacIsaac 1986.

23. "Brinkmanship is thus the deliberate creation of a recognizable risk of war, a risk that one does not completely control. It is the tactic of deliberately letting the situation get somewhat out of hand, just because its being out of hand may be intolerable to the other party and force his accommodation." Schelling 1960, 200.

24. This argument was based on deterrence in a prevalently bipolar situation. Problems of stability-instability in a multipolar or even post-proliferation world will be addressed in chapter seven.

25. I am here using "strategic" in the original meaning (cf. Freedman 2003, 5) of an independent (from other services) use of air power, targeting the economic and social structure of the enemy as well as his morale; an undertaking which is a legitimate offspring of the total war concept. Otherwise "strategic" is used according to its merely quantitative meaning in nuclear jargon (long-range, intercontinental arm systems).

26. Both terms are used by John Mueller, quoted by Kaysen 1991, 82. I am not going to explicitly discuss Mueller's thesis asserting the irrelevance of nuclear weapons nor his critics' remarks, since my stand has been sufficiently clarified in this chapter.

27. This definition reproduces Robert Dahl's definition of influence, Dahl 1963; its adoption as a definition of power has been suggested by Bobbio 1981. Since it includes the possibility that power breaks up not just the resistance, but the inactivity or apathy of the other actors as well, it is more general than Max Weber's notion of power in general (power/*Macht* as broader than domination/*Herrschaft*), cf. Weber 1922, 53. A more differentiated discussion of "power" is in Hill 2003, 129–38.

28. War may seem to be identical with compellence, but pre-emptive war is not, and neither is the side effect, the deterring message that the war waged by A against B is intended to send to C or D, the possible allies of B. On the difference between threats that compel and those that deter cf. Schelling 1960, 195–99.

29. Gunpowder—as we can read in the lectures that Hegel gave at the University of Berlin between 1822 and his death in 1831—turned war from being a contest based on the "particular physical violence" of the fighter's arm to a fight based on the rational qualities of the leadership. "Firing is aimed at the abstract, general enemy," fighting is no longer driven by hatred for the individual enemy (Hegel 1968, 855–56, my translation).

30. On acceleration in total and nuclear war cf. Virilio's (1986) philosophical account.

31. Even if not an actual concern, the American nuclear umbrella still seems to shield the EU countries, in spite of their wide dissent (UK and Poland aside) from Washington's foreign policy in the first decade of our century.

32. In some corners of the antinuclear movement of the early eighties in Western Europe this issue used to be addressed under the slogan of reinstating national sovereignty as constitutionally guaranteed popular sovereignty, thus rejecting NATO's decision (1979) to deploy the Euromissiles or get out of NATO altogether. To ask that Western Europeans and Canadians be entitled to vote in the election of the U.S. President, the statesman who makes the ultimate decision whether to use nuclear weapons, would have been a less regressive and more realistic proposal.

33. This behavior—unlike the exit behavior by fear which I shall discuss shortly—is still rational, or can be reconstructed as such, even in terms of instrumental rationality: one should not invest time and energy in undertakings, like political life in the nuclear state, which do not reward one with the expected benefit, i.e. security. Equally rational, but hardly referable to the instrumental type of rationality, is the "voice" behavior of those, e.g. in peace movements, who explicitly request the state to provide a credible security. "Voice" and "exit" are of course used according to Hirschman 1970.

34. Common security as a policy was launched by the Palme Committee in 1980–82, and theoretically refined primarily in Germany by a research group led by the SPD expert and policy-maker (*Ostpolitik*) Egon Bahr, cf. Bahr and Lutz 1986 and Ragionieri 1989. Freedman (2003) speaks of "mutual assured security", 414–18.

35. In the explanation of the nuclear arms race during the Cold War, this mental attitude is prior to other, more visible factors like the influence of the military-industrial complex in the USA or its counterpart in the Soviet Union. This influence could only be effective against that cultural background and was not the primary source of the arms race, cf. Weart 1988. On apocalyptic elements in that background cf. Weber 1997, chapter 11.

36. I am using "reification" in the Marxian meaning: social or political relations between human beings and groups seen as if they were relations between *res*, things. Fetishism is the other side of the coin: things are believed as such to have human character and exert power on human beings. Reification processes need to be unveiled, whereas a fetish—missiles or warheads as tokens of security—is more evident.

37. The country which for longest resisted its own reshaping in order to accommodate new military or industrial techniques lost power and (relative) wealth to its speedier competitors; domestically, the longer a country needed to adapt its institutions to the Industrial Revolution, the longer-lasting and more incisive the costs paid by the working class and peasantry.

38. See above § 1 in this chapter.

39. A prime example was the failure of the (Oppenheimer and) Baruch plan of 1946 due to the Soviet lack of interest in anything short of their own bomb and uncertainties in the American position (cf. Bundy 1988, chapter iv).

40. One of the problems of investigating these phenomena is that they can hardly be referred to just one aspect—nuclear threat—of the environment from which escape or salvation is sought. Lack of human contacts beyond the labor and administrative sphere, the crisis in value and authority patterns in the reproduction of society through family and education are quite inextricably intertwined with the impact of global threats.

41. More on this in chapter six.

42. That is the rational, i.e. optimal, choice of a means to an end in the interaction with at least one other actor. It also includes instrumental rationality, which considers how to employ things without the intervention of other actors.

43. "It was difficult to attach any rationality whatsoever to the initiation of a chain of events that could well end in the utter devastation of one's own society (even

assuming indifference to the fate of the enemy society.)" This is how far Freedman (2003, 461) went in recapitulating the history of nuclear strategy.

44. On unity and fragmentation of the notion of reason in modernity the best account is still to be found in Habermas' *Theory of Communicative Action* (1984–87). More on rationality at the end of chapter six.

45. See chapter five, § 4.

46. The previous assessment of the environmental impact to be expected from the use of certain materials rather than others in public works or house building goes in this direction.

47. The Test Ban Treaty and Non-Proliferation Treaty were stipulated as late as in the sixties, nearly twenty years after the bombing of Hiroshima and Nagasaki and only *after* the nuclear saber rattling of the Cuban missile crisis.

48. I borrow "the joy of insight" from the title of Viktor Weisskopf's autobiography (1991), while "the trial of theoretical curiosity" is the second part of Hans Blumenberg's seminal work (1983) on the historical meaning of modernity.

Chapter Four

The Challenge of Global Warming

The recent warming of the earth that has come to be known as global warming is more exactly an increase by nearly 0.7 K during the twentieth century in the average temperature of the atmosphere surrounding the globe; this is clearly a statistical entity,[1] and by far not a nearly evenly distributed phenomenon. Little increase has been for example registered so far in the Antarctic, while the rate of temperature increase is less relevant in the equatorial and tropical areas. But what matters more is the strongly different impact of warming temperature on the various countries, economies, and cultures according to their individual capacity to adapt to the new conditions and to shelve themselves from disrupting effects. Wealthy societies are better off than the underdeveloped ones. More desertification in Africa hits men, women, and children far worse than exceptionally hot summers in countries in which water supply is not a problem and fans or air-conditioning are easily available. Beyond being inaccurate in social analysis, "global warming" is, if taken literally, an expression lacking fairness, as it may falsely suggest that everybody on the earth is under the same duress because of it. In recent years natural as well as social scientists have tended to replace this term with "climate change," which is indeed more precise and comprehensive. Yet in this book we are not looking at this natural phenomenon as such, but rather at its impact on humans and, what is more relevant, its being a result of human agency. We should then qualify the expression used and speak of "man-made" or "anthropogenic climate change," but this is admittedly a pretty cumbersome wording, which cannot be used everywhere. This is why in the following I shall use all of the expressions mentioned in this paragraph (global warming, climate change, and man-made climate change) as if they were equivalent, and as required by the argumentative context and without attributing a particular edge to any of them.

In this chapter I shall first examine the reasons why climate change can or cannot be considered a truly global challenge (§ 1) and consequently look into the strategic and moral complexity of the action patterns it may suggest (§ 2). Does it generate conflicts, and which type of conflict? Why do efforts aiming at protecting the atmosphere sometime work and sometime fail? Institutional as well as moral and subjective factors come into play, which I shall discuss in § 3. The deep reasons for failure are discussed in § 4 under the headings of a "Promethean gap," a "tragedy of the commons" and a tragedy of democracy. How all this could come about is the question raised in § 5, in which the dialectics of the modern approach to nature is outlined.

The subject matter, the type and style of knowledge, and the language are in this chapter fairly different from the two previous chapters. The question of how far nuclear weapons and man-made climate change raise similar problems and allow for joint or parallel philosophical considerations will emerge occasionally in this chapter and be addressed in the following ones.

1. HOW GLOBAL IS MAN-MADE CLIMATE CHANGE?

Climate change has always existed, as have global warming and global cooling and, needless to say, greenhouse gases, the very pre-condition for life to be possible on this planet.[2] By "climate change" we presently mean more specifically a set of phenomena, such as global average air and ocean temperature rise as well as rise of the sea level, precipitation change, droughts, and floods that can impinge on essential aspects of our life on earth: food and water resources, ecosystem and biodiversity, human settlements, and human health. There is now enough evidence, as we shall soon see, that much of the warming observed particularly over the last fifty years is attributable to human activities, particularly fossil fuel use and land use change, solar variation and sulfate aerosols accounting for the rest.[3] In this I am strictly following the IPCC Third Assessment Report (IPCC 2001, 3), or the first part (out of three) of the Fourth Assessment of 2007, which deals with the physics of climate change and was released shortly before this book went to press. I am relying on these authoritative documents of the UN-sponsored Intergovernmental Panel on Climate Change, a highly institutionalized epistemic community, because they strike a reasonable and publicly tested balance between different views of the present situation, the causal explanations, and the expected developments. Where no decisive and definitive scientific evidence and explanation are available,[4] which is often the case in a new science like climatology, only a "negotiated assessment" (Young 2002) is possible, best if reached by transparent, broad-based, and public debate like in the IPCC case, a circumstance that somehow resonates with the consensus theory of truth debated in philosophy in the last decades. On this ground and in the context of this book I do not see the necessity to debate the harsh, but mostly poorly argued criticism (see however note 3) which in some corner has been brought against the theory of global warming now prevailing in the worldwide scientific community.[5]

A first real challenge for humankind lies in what may happen in the near future, that is the time from now to 2100, to which most of the IPCC forecasts are limited. For all of the socio-economic scenarios taken into consideration and the resulting emissions (disregarding Kyoto or other possible cuts), in 2090–2099 (as relative to 1980–1999) temperature is expected to rise by 1.8° to 4° and the sea level by 0.18 to 0.59 meter (IPCC 2007, Table SPM-3, 13). The five hundred scientists who contributed to the IPCC Fourth Assessment Report regard as very likely (90% chance to be true) that the steep rise in warming observed over the last fifty years is due to the increase of greenhouse gases concentration in the atmosphere, particularly of CO_2, which in 2005 has reached 379 parts per cubic meter (ppm) (IPCC 2007, 2),[6] compared with about 280ppm in the pre-industrial era; not to consider the contribution of other anthropogenic greenhouse gases such as methane, nitrous oxide, and tropospheric ozone. The larger the changes and the rate of change, the more the adverse effects will predominate over the benefits of warming up the colder areas of the earth, particularly with regard to tropical and subtropical regions, where the cereal crop yield will decrease and water shortage will affect more people and contribute to more heat stress and infectious disease epidemics; on the other hand crop yields will increase in Siberia and Canada and cold-related morbidity and mortality decrease, though this is counterbalanced by the soil erosion brought about by the thawing of permafrost. As to the rising sea level, low-level coastal regions and small islands will be inundated, destroying agricultural land and driving the local population to migration. In a word, even if the physics of climate change is a new field still in need of much development, there are enough serious signs of a link between humankind's activities and the speeding of global climate change as to justify taking action for the philosophical reasons that will be discussed within the next chapters. I am not going to additionally investigate the reasons for taking action against local warming and pollution, as even in the U.S. states such as California and the New England commonwealths have done for the past several years.

So far we have examined only the most relevant changes and their effects as shown by the projection of present conditions to the time up to 2100. Not all of these predictions have the same level of probability in the IPCC assessments and only those with a high level have been quoted here. Looking "beyond 2100" becomes more uncertain, as little is predictable when it comes to the possibility of abrupt non-linear changes in many ecosystems, such as large melting, even if in different times, of the Antarctic and the Greenland ice sheet, and the weakening of the thermohaline circulation of the oceans (cf. IPCC 2001, 14–17), which according to an hypothesis now under review could even deflect the Gulf Stream away from Europe. These changes would be likely to be irreversible, forever upsetting at least in a wide portion of the globe the geographical and environmental conditions that have made possible and dominated the birth and the spreading of civilization. All this is not only much more severe and at the same time uncertain than predictions limited to the twenty-first century; it also regards generations of a future farther away, to which we may feel less responsible than to our children, grandchildren, and great-grandchildren—a problem that we will have to come back to when re-considering these predictions under the moral point of view.

Now, our first problem is that the projection of global warming into the near future shows that it is much less global than its name pretends, while its truly global aspect (the potential to directly and severely affect everybody on earth) regards generations who will live on the planet in hundreds or thousands of years from now; the prediction itself has a greater uncertainty than for the nearer future. These conflicting statements need further explanation.

As to climate change until 2100 (let us call it CC1), it may make life less pleasant or more risky to many people everywhere,[7] but it will hit people in developing countries and poor people in all countries in a disproportionately severe way (IPCC 2001, 12), particularly because of water shortages and decreasing yields of cereal crops in most tropical and subtropical regions. On a global level climate change is not going to significantly affect food security for the next decades, but locally substantial negative effects are expected for developing and low-income countries; by 2050 two hundred million people may be permanently displaced by flood and drought in several regions of the world (*Stern Review* 2006, vi). This means that for the current century global warming cannot be expected to become a cause of immediate damage, increasing danger, and overwhelming concern for a majority of the world population; it is unlikely to foster across countries and continents a will to protect ourselves that may easily overcome national reservations and create a centralized mechanism of decision-making endowed with the legitimacy of a survival-providing power. *Caeteris paribus,* efforts aiming at the protection of the biosphere[8] will remain a matter of conflicting opinions as to how far emissions generate climate change, how to meet the problem (by mitigation or adaptation,[9] or by which mix of them?) without putting economic stability in danger and how the burdens should be distributed among developed and developing countries. Needless to say that these problems and processes remain confined by their very nature to the intergovernmental terrain and do not seem to pave the way for an emerging superior power, unless this is established for other reasons (the nuclear threat, for instance) and by other pathways (an imperial power successfully asserting itself, or a free world federation of Kantian ascent, to mention the two extremes, unlikely that they may be).

Things would change if the medium-to-most–unfavorable predictions for the centuries after 2100 should prove true and the warming become really global (CC2, the climate change that may happen from 2100 on); in one word, if Manhattan starts getting under water because of the melting of Greenland ice. This is a cynical, but realistic remark: it will be, if any, the flooding of Times Square rather than of the Bangladeshi coastline that will prod the media networks and all politicians to proclaim climate change an urgent global threat. Concern and fear, even panic, would then become the driving motivations not just among the public, but the decision-makers as well. Economic growth or lifestyle[10] priorities would range behind the necessary changes in technology, economy, and social behavior in the individual countries, and a worldwide consensus would be easily found on shared environmental standards; they might even be imposed on recalcitrant countries by economic or military coercion. This would be a survival scenario, in which global warming would claim attention as a fully global challenge, both universal and lethal. This scenario cannot be discarded as mere catastrophism, because there is already enough scientific evidence for it becoming possible (possible, not highly likely) and enough consensus in the scientific community on this possibility.[11] It will have to be taken into account later, when arguing the normative and political consequences of CC2, or rather the different consequences of CC1 and CC2. But scientific predictions must remain inconclusive as long as more sophisticated models of climate evolution[12] across the ages are not established. Science is presently not able to confront us with an unequivocal image of threats that we all feel compelled to counter by the most reliable course of action we have the power to take. As is often the case, what to do or not to do, or how long to wait before taking action, remains a matter of opinion in the scientific estimate as well as matter of choice in the moral and political field. Still in environmental policy-making, images of doom, projections of fear and hope, and similar subjective attitudes co-determine our course of action along with rational analysis and calculation.

For the first segment of foreseeable climate change (CC1) until 2100 our knowledge seems to be more reliable as far as estimates of change and assumptions about the causal link with human activities are concerned. But in this segment of time global warming, as we have seen, has turned out not to be truly global in the distribution of the damages it generates. Let us now re-assess the legitimacy of its claim to be global and to deserve the same consideration given to the nuclear threat.

Global warming (CC1) is *not global* because it is unevenly distributed:

- geographically (low-level countries, small islands, areas prone to desertification); and/or
- socially (the poor who are the majority in most of those regions as well as the poor and the less-wealthy elderly in all countries are more severely hit).

Global warming remains *global* under three aspects when observed from outside:

- the systemic dimension, as it is understandable only on the basis of physical processes regarding the entire planet;
- the causal as well as moral (responsibility) link that binds together perpetrators and victims (many of whom can under certain conditions become victims and perpetrators themselves), e.g. the people living in an air-conditioned building in Chicago or Berlin and those fleeing the flooded coastal villages of Bangladesh;
- the fate of shared agency, an essential feature of global challenges: only acting together cooperatively can the problem be addressed, while there will be no major (or not even a significant) mitigation of greenhouse emissions as long as today the USA and tomorrow, say, India stay away from an international agreement.

Besides, global warming seen from inside, that is as it is perceived by participants (social and political actors), can be dealt with as a global threat if

a. the better-off countries regard disruption of social life in developing countries due to climate change (epidemics, famine, mass migration) as a threat to global security;
b. considerations of fairness come to play an important role in international decision-making, by the force either
b1. of public opinion in most of the developed countries ("we are the main polluters, we have to take action and to pay for it") or of
b2. a new international balance of forces, in which developing countries successfully claim the right to be protected against further climatic damage and not to have to sacrifice their growth altogether.

Whether a. and b2. will occur depends on physical and political events and the way they may be interpreted in strategic terms by the actors as utility maximizers. The shift to taking global warming more seriously would insofar bring no novelty as to the method by which actors may get to it: it is usual that a more sharply perceived threat or more pressure exerted by competing actors drive them to rethink their course of action. Novel would be only the b1. case, because fair distribution of burdens would be chosen on behalf of fairness itself, as required by environmental movements, and not for strategic reasons. These are obviously ideal types of action and motivation, none of which comes up in reality without some mixing with the others. That is why it is important to keep debating in the public sphere as well as trying to orientate public opinion along universalistic criteria of fairness and to bring them to bear on political decision-makers. But a universalistic sense of fairness cannot really be credited to substantially change the course of action taken by countries that are presently on the winning side in the uneven distribution of damage and suffering attributable to climate change. When they, like the European Union's member states, have decided to stick by the Kyoto Protocol out of enlightened self-interest and also because they have to pay only low costs,[13] fairness considerations can only reinforce this decision and help sell it to the public. Where, as in the United States, the interpretation of the current "national interest" leads elsewhere, it is hard to see how a far-sighted threat perception and enhanced environmental fairness can influence the American public opinion beyond what, Al Gore and Arnold Schwarzenegger notwithstanding, remains its (liberal) minority. Admittedly this is relevant in political, not in normative terms, but the political difficulties of getting the fairness-oriented solution through reverberate on the normative approach, as we will see.

In any case, fairness, though maintaining its universalistic premises, is unlikely to change policies if it does not strike a balance with *effectiveness* (as a tool of a more comprehensive fairness) and with a prudential approach, in other words if it does not help enlighten our self-interest. To make the polluters pay while exempting the polluted countries of the Third World may sound the top of fairness, but does little to reduce global warming and to prevent our posterity from being damaged more severely than we are.[14] Fairness against future generations must come to terms with fairness towards those who are polluted and poor today. The previous huge imbalance between the greenhouse gases emitted in developed and developing countries will come to an end around 2020, when the latter will send more CO_2 into the atmosphere than the former; but exemptions in emissions reduction agreements are already now self-defeating, also because of the inertia effect. What is this?

2. A COMPLEX ACTION PATTERN

According to the IPCC Third Assessment, inertia means that some effects of anthropogenic climate change may become apparent at a slow pace and draw our attention only when it is too late to do something effective; but it means as well that the greenhouse effects of emissions will keep their current level well beyond the time in which we may have stabilized the emissions. Even if CO_2 emissions were reduced to a small fraction of the current level (a target that widely exceeds Kyoto's), it will take 100 to 300 years for the atmospheric carbon dioxide level to stabilize and a few centuries for temperature stabilization; but to compensate the effects of sea-level rise due to thermal expansion will take centuries to millennia, and several millennia in the case of the sea-level rise due to ice melting. It cannot be excluded and it is in some cases plausible that in the course of time those effects cross "associated thresholds" (IPCC 2001, 16; see also Figure SPM-5,17), thus making changes like those in the ocean circulation irreversible.[15] This makes things more awkward also from the philosophical as well as the specifically ethical point of view.

Philosophically, the mere possibility that by the force of accumulated greenhouse gases and their persisting effects some catastrophic threshold (say, the melting of a fair amount of Greenland ice with the subsequent submersion of large coastal areas and possibly the unraveling of the equilibrium based on the Gulf Stream) is already going to be inevitably and irreversibly reached adds a sarcastic, if not tragic, undertone to the present debate on pros and cons of mitigation and adaptation policies. At the surface this debate is about how to optimize choices between some more warming *versus* some more costs and involves actors with sufficient, though not complete information. Should the inertia effects have meanwhile and without our knowledge doomed all efforts to fix global warming, it would be a debate on whether or not to repaint the façade of a building that will collapse in six months because we have put a gas pipe underneath it and pumped gas, then forgetting about it. The focus is here on the disproportion between the strategic rationality that applies to business-as-usual and the (lacking) consciousness that we may have already upset our planet—unwittingly, or rather with the pretension of reshaping it in a more comfortable way for our needs and wits. The philosophy of the perverse effects of humankind's handling of nature cannot blow out considerations of strategic rationality, but it could perhaps give the debating actors a sharper sense of the uncertain limits of strategic reasoning.

In ethical terms the *inertia phenomenon* burdens on us a still greater responsibility towards future generations (granted we recognize any such responsibility, a fundamental problem which shall be discussed later in this book). If we live in an environment threatened by possibly irreversible inertial processes, we are responsible for not worsening the life conditions of those who will later dwell on a planet that will be more endangered than it is for us, because the amount of greenhouse gases in the atmosphere and their inertial impact will be in any case greater than it is in our time. Hence arises to us an obligation to decrease emissions as far as it is compatible with the preservation not of our "life style," but of acceptable standards of civilization and consumption, which entails economic growth, but not growth of any kind and at any cost. Further enhancement of our knowledge about climate change and especially the role of inertia may justify strengthening or abating the precautionary measures the present stand of knowledge leads us to take. The "precautionary principle" is a wide and (because of its costs) problematic ethical matter, which shall not be examined in this book. For our inquiry suffice it to say that precaution in matters of global challenges establishes an obligation to act for the care of future generations with measures whose volume may seem to be disproportionate to the present degree of certainty about the threat level. In other words, if our doing nothing or little can endanger their elementary life conditions, we are justified by the dimension of what might be at stake if we take those measures. Even if there is no certainty, the simple likelihood of a very large human suffering is reason enough to be cautious, as we have seen in the risk chapter (one) of this book. The need to learn more and not draw wholesome and hurried conclusions is one thing, cheap skepticism on present knowledge versus reasonable preoccupation with it is another.[16]

A last element can be added to this pattern of strategic as well as ethical complexity. Rethinking global warming as a global challenge implies some correction in the description of an actors' type that frequently shows up in the literature: the *free rider* in international regulations concerning environmental problems.[17] Free riding can mean either non-participation to an agreement or regime or, when participating, non-compliance with the standards and the deadlines agreed upon among partners.[18] But actors that currently or in the future have a large amount of greenhouse gases emissions or ozone-depleting chlorofluorocarbons release are not in a position to properly free ride, as by definition this behavior would mean to receive the benefits of collective action without paying one's own share of costs. By attempting to free ride, that is, to not participate or to skip their obligations to reduce emissions, major polluters do however largely spoil the collective endeavor to stabilize CO_2 levels, or at least they delay the attainment of the intended targets, thus enhancing the risk of significant thresholds being crossed by the effects of inertia before any intended benefit materializes. In this case, even for the would-be free rider such as the United States during the Bush administration, benefits of collective action do not accrue at all or come up much later, perhaps too late, if it is true that we have already put a time bomb on our planet and live on borrowed time. It is only smaller countries whose emissions do not substantially alter the global emissions level that can be free riders successfully and in the proper sense.

3. POLITICAL UNCERTAINTIES

As we have seen at several points of our brief review of climate change, this set of phenomena raises philosophical, normative (legal as well as moral) and political questions. It is now time to step further towards a systematic account of these questions. But philosophical (say, the approach to nature in modernity) and normative (say, obligations to future generations) issues implied in climate change are largely, although not totally, identical with those posed by nuclear weapons; they are therefore better dealt with in a general overarching argument (see chapter five and six). On the contrary, political questions, i.e. questions concerned with conflicts and power relationships, are obviously as peculiar of global warming or ozone layer depletion as are the physical damages and the policy responses generated by them; hence, they shall be discussed in this section.

While the threat represented by nuclear weapons is itself the result of a pre-existing conflict (first between the Allies and Nazi Germany, then between the Soviet Union and the United States of America), climate change and its consequences are rather generators of conflict and not the result of it.[19] This is not something new in history,[20] the novelty resides rather in the prevailing anthropic genesis of the change as well as in the fact that the related conflicts are now contributing to restructure the political agenda worldwide. As it is primarily its conflict-generating potential that makes man-made climate change political, it is perhaps useful to classify it among the several types of environmental conflicts, a short for "conflicts regarding renewable resources located in our physical environment and becoming scarce or globally endangered" and meant to be civil or interstate/interethnic or global.[21]

Renewable resources can be divided into four groups, as far as the structure of conflict is considered.

1. Resources which are subject to adversary and unequal *distribution* between groups: conflicts regarding them tend to be zero-sum games: cropland, drinkable water, fish stocks, forests as a source of wood. Depletion caused by desertification, pollution, unregulated fishing, and logging leads to distribution conflicts between competing groups which can (or believe they can) only ensure their own survival by winning a larger share of those resources. This conflict is about redistribution, not retribution or compen-

sation: the fact that desertification is fostered by global warming, which has its main source in the emissions from developed countries, is not perceived as an element of strife. This conflict is typical of developing countries and can be classified as a competition between countries, ethnicities, and tribes, for example in the Sahel region.

2. Resources which are indivisible by their very nature, but local in range (also called *local commons*, or local public goods): clean air (as well as forests as an element of its regeneration), non-polluted water in rivers, lakes, and seas as an element of the overall local environment and the like.

   This is the usual origin of domestic environmental conflict, and can have the same effect in the international area wherever a free public opinion and an educated political elite have grown sensitive to troubles of this kind. A certain degree of democracy seems to be a presupposition for this conflict. It is not specific to the North or the South of the world and can arise everywhere, but its political precondition is more likely to be found in the North or more precisely in the West. Examples are the recurrent German demonstrations against nuclear waste being transferred from France to German deposits.

3. Resources which are indivisible by their very nature and truly global, that is, they are equally necessary to the whole humankind, while their depletion affects everybody on earth: *global commons.*

   Examples are clean air as far as its pollution may result from worldwide processes, stable temperature as an element of climate stability (i.e. the opposite of global warming), preservation of the stratospheric ozone layer as a protection against UV-radiations and resulting warming and diseases. Conflicts originated by the depletion of these resources require a high degree of sensitivity for general issues, that is, for those not related to special-interest-groups. In other words, they require not just democracy as a basis for debate and inquiry, but even going beyond democracy, or rather beyond the present entanglement of the democratic method of conflict resolution with self-centered interest groups, which hardly have any encompassing view of general and future problems. International cooperation as the only effective way of coming to terms with these phenomena thus collides with the belief that they are not relevant or can be managed by a country on its own exclusive terms (adaptation vs. cooperation, as Soroos put it). This is a typical North-North conflict, the Kyoto Protocol being its most striking example.

4. A last type of conflict can arise (in the future presumably much more than in the present) if the South requests compensation for environmental depletion such as desertification, blaming the North as its main culprit; whether or not this is true is a matter of public belief rather than scientific

evidence. The provision of the Kyoto Protocol which sets for developing countries much lower targets in emission reduction goes in this direction—an issue of both cooperation (the Protocol) and conflict (the Bush administration's withdrawal from the agreement made explicit reference to this provision). The rationale for identifying this kind of conflict lies in the recognition that damages to the global engagement, albeit they are able to hurt everybody on the earth, do indeed hit the poor and less developed populations harder than affluent societies, which are to a limited extent protected by better technologies and the social network of disaster relief.

This typology should help us to uphold the distinction between global conflicts, or conflicts around global challenges (global warming and ozone layer depletion) such as in type three and four, and other environmental conflicts, whose actors and stakes are pretty different.[22]

A further step is to look at the specific attempts to settle or to prevent those conflicts, although the problem is not simply conflict-solving, conflict-resolution, or prevention, but effectively meeting the challenges as well: no outcome does in itself guarantee the coincidence of the two objectives. They are both political, if it is true that *politics* is not just about winning/losing or peacefully and cooperatively settling conflicts between actual parties, but providing effective protection and conditions of well-being for present and future members of the polity. Now, the present situation is a mixed one: one target (reduction of CFC release in order to restore the protective ozone layer) seems to be substantially met, even if the restoration will still take about sixty years because of the inertia effect.[23] On the other hand the Kyoto agreement on the reduction of greenhouse gases emissions has lost efficacy because it was not able to involve the historical as well as the upcoming big polluters. It is also interesting to look at this difference between the two processes from the epistemological viewpoint of competing theories of politics, especially of international relations.

What has been significantly achieved for the effective protection of the environment on international or planetary scale has been achieved through intergovernmental cooperation, not through the mere unplanned intersecting of actions taken by single states for the protection of their own citizens (the realist way); nor seems the imminent emergence of a supra-national authority designing and enforcing rules on the several actors (as cosmopolitans wish) to be a reliable assumption. It comes as no surprise that the most encompassing and authoritative political science literature on environmental protection across borders is informed by the neoinstitutionalist approach.[24] It maintains that the outcome of intergovernmental agreements or, more generally, the institutionalization of international relations is different from the algebraic sum of single states acting on self-interest, because institutions (i.e. shared sets of rules, whether or not formalized) modify expectations and incentives, thus facilitating self-restraint over confrontation, cooperation[25] over non-participation or free riding. We cannot rerun here the neoinstitutionalist argument against neorealism,[26] which would lead us far away from our specific subject matter; there is little realist literature on environmental issues and policies, so that the discussion would have to be run on general theoretical terms. In any case, the neoinstitutionalist account of international governance on environmental issues is not the last word on this matter, and this not only because other conceptual frames of reference such as reflectivism and/or constructivism (cf. Paterson 1996) have been brought to bear. It is the very political experience of intergovernmental cooperation that lies in the background of neoinstitutionalism that has become questionable as far as it seeks to be seen as an ongoing and reassuring path towards effective governance of global warming.[27] In this direction, it shall be instructive to look at the explanations given to the diverging outcomes of Montreal 1987 and Kyoto 1997.

As a premise, we must note that "institutions for the earth" only work effectively under the political circumstances that the editors of the volume devoted to them (Keohane et al. 1993) phrase as high level of governmental concern, hospitable contractual environment, and sufficient political and administrative capacity of the national governments involved (vice versa, established institutions help to maintain and enhance those circumstances).[28] Other circumstances of economic nature must be present and were present at Montreal 1987: a favorable cost-benefit ratio,[29] the twin (stick and carrot) possibility of imposing damaging and deterring sanctions (trade restrictions) on non-compliant actors, while legitimizing this threat by side-payments pledged by wealthier partners to the weakest economies to facilitate their adaptation to the new standards. This is not the case with the Kyoto Protocol, in which the global marginal benefits are expected to exceed every nation's own marginal benefits (Aldy et al. 2003, 6). Let us however remind that the cost-benefit calculation is not the last word about the criterion as to whether or not to participate in environmental agreements, because "nations,"[30] or rather political regimes and governing coalitions, are also motivated by cultural differences as to how to assess the value of global *vs.* particular benefits, especially when these are projected into the future.[31] But it is true that cost/benefit considerations play a central role anyway and can even be a stumbling block as long as incentives are not restructured, and/or cultural change has not moved environmental protection up the value scale which presides over political decision-making in democratic regimes, overcoming what in chapter five shall be discussed in terms of "generational nepotism." Coming finally down to the economic aspects of Montreal and Kyoto, a further reason for the success of the former has been the limited and well-defined amount of change to be made (replace CFC by HFC and HCFC, which are themselves greenhouse gases, to be phased out by 2030 under the Copenhagen Amendments of 1992, cf. Barrett 2003, 222–37) by employing available technical resources; while Kyoto requires a vast amount of new technologies not yet available on a large scale, or not enough tested,[32] and its costs in terms of decreasing GDP (1% each year) are to be taken seriously—even if GDP increases should not be relished as an ultimate fetish and rather combined with other value-indicators such as the Human Development Index. In any case, the availability of new technologies as well as the possibility to promote by effective treaty provisions their transfer and their spreading to all polluting countries remains a pivotal element of successful intergovernmental policies creating new regulatory frameworks for economic and cultural change in the right direction. As to the restructuring of incentives, this notion can be illustrated with two recent examples: if the "greening of America" signaled by *The Economist* on 25 January 2007 is to be taken seriously, a major motive for reducing its dependency on carbon fuel by shifting to ethanol and other energy sources appears now to be the political will to alleviate its dependency on the oil-producing Islamic countries and Chavez's Venezuela. A typical neoconservative attitude such as the priority of (economic, military) national security seems now to reap green fruits, which were altogether unthinkable in the neocons' heydays. The other example is the important attempt made by Tony Blair's cabinet and particularly by its chief economist Sir Nicholas Stern (2006) to shift the attention from the costs of emissions reduction (cf. Nordhaus and Boyer 2000) to the dramatic economic costs (0.5–1% GDP yearly) that are to be borne in the future in case no action is taken right now as to contain the damages caused by man-made climate change. This causes and will keep causing actual economic losses to everybody rather than human suffering to the disadvantaged alone.

To work effectively, environmental protection on the global scale needs however other favorable circumstances beyond the political and economic ones. They may be called cognitive and motivational: the problem to address must come up with enough scientific evidence, and it must affect public opinion and decision-makers in a clear and challenging way. The first standard was met in the case of the ozone layer depletion,[33] also because the causal chain (from the use of CFCs to the catalytic reactions destroying the protective O_3 layer and further to skin cancer due to unfiltered UV-radiations) is relatively clear and straightforward. This cannot be said of the cause-effect link that may exist between expanding fossil fuel consumption and greenhouse gases emissions and the global increase of the planet's average temperature; as noted before, we even lack until now overarching and generally accepted models of climate change across the ages and particularly of oceans-atmosphere circulation. Making the reduction of greenhouse gases emissions a policy issue is reasonable and necessary, because there is evidence enough to be worried about, but it requires a more complex reasoning (including the moral aspects mentioned in this chapter and discussed in chapter five and six) than the manifest "Montreal" sequence of "damage revealed (the depletion) / damage's causes (CFCs) and effects (skin cancer, primarily but not alone) identified / causes removed and replaced with other means (HFCs-HCFCs) without net losses of economic welfare."[34]

Moreover, the transfer of knowledge from the cognitive/scientific level to the motivational field of public opinion's attitude and political agenda setting is not guaranteed and can be delayed or distorted, but sometimes accelerated in an unexpected way.[35] This has to do with the complex and widely manipulated way how themes are chosen for public concern or dropped from it, images and symbols created or rapidly dissolved in the present universe of global communication. But it is also influenced by shifting majorities in interest and government coalitions, which may prevent a scientifically well-established alert notice from becoming a policy issue, or even contest treaty obligations when the domestic coalition comes under the sway of a party or lobby that regards itself as a loser in the treaty trade-offs.

Other motivational features result from the ethical terrain as well as from the threat perception. The Kyoto policy of imposing emissions reduction targets only on the developed countries out of fairness towards the developing countries that pollute in a hugely less amount, but are equally subject to the greenhouse effect caused by the former does not seem to strike a balance between fairness and efficacy: the emissions in the developing countries, already amounting to 21 percent of the total, at the signing of the Kyoto agreement, will significantly increase in the next decades inasmuch as huge economies such as China or Brazil will hit higher development targets, sometimes also using older and more polluting technologies or sources of energy. The fairness-motivated compensation for those countries entering the international effort against global warming should be found in monetary terms or technological aid rather than in dispensing them from target obligations. This would enhance the legitimacy of an international agreement in the eyes of the public opinion in developed countries and neutralize the opposition of those who are impermeable to whatever fairness considerations.

Finally, the public perception of the specific threat must be dramatic and widespread enough as to generate requests of change addressed to decision-makers and to provide legitimacy for new measures against the opposition of particular interest groups. Ironically for those who believe that security problems are a perversion of politics (by themselves and not only in certain circumstances), the "securitization" of climate change helps provide enough motivation as to move things around. Under this regard, the threat carried by global warming looks pale if compared with the nuclear war's. For the lucky circumstance that the generations living after 1945 were spared nuclear destruction we have to be grateful to and always reminiscent of the dead in Hiroshima and Nagasaki. Without the worldwide known pictures of carnage that have shaped the imagery of all human beings who went to school somewhere for the past half century it would have been perhaps impossible to motivate the ruling elites to be cautious in front of the nonetheless existing nuclear weapons, which have been employed for deterrence and not for war and whose arsenals have been later reduced. On the contrary, there is so far no picture and no account of global warming that can hit the imagination of ordinary citizens in a politically dramatic way. The few who have direct experience of the related pain, the people suffering from increased desertification or coastal flooding, are speechless pariahs in the present sophisticated world of political communication. As noted before, global warming is not or not yet truly global, a circumstance that does not help to build the sense of community which can "function as mechanism of social control among users of Common-Pool Resources."[36] Climate change is a process connecting many events none of which is defining, while most are by now just micro-events (temperature changes, sea level increases) whose significance is mostly statistical and among the present generations can be grasped at best by farsighted elites, not by the average voter. An impressing shift in the threat perception is needed also because the present generations seem to be much more keenly aware of the economic cost of reshuffling production and consumption[37] than of the threat itself in its twin aspects, the sufferance impending on future dwellers of the planet and the even higher costs they will have to pay for containing it (adaptation costs). The time dislocation, partly based on the physical inertia effect, represents a further relevant difficulty: the economic costs have to be paid by the present and next generations, the benefits will accrue only to the far future ones. A society has to be ethically convinced of the obligation it has towards them, as it cannot be a strategically managed sense of self-interest that pushes it on the road to reduce greenhouse emissions. Even the refocusing of our attention on the present and particularly future costs of global warming attempted by the Stern report will work only if we will feel a non-economic tie, that is solidarity, with our posterity, which makes us feel their economic losses as if they were our losses.

Which conclusions can we draw? As a matter of fact, only international cooperation in intergovernmental institutions, not a non-existing world government or parliament, perhaps not even a worldwide Environmental Protection Agency (see chapter seven on this), seems so far to be able to address effectively environmental global challenges such as the warming of the atmosphere and ozone layer depletion. But this only existing level of action cannot guarantee that measures are taken, and, if taken, taken in time and in the appropriate scale, let alone that these measures are timely and really able to solve the problem.

We have gone through some of the reasons that make the work of international cooperation not fully reliable in terms of a satisfactory problem-solving. They are the "normal" reasons that impede, delay, derail, and in any case transform political processes of decision-making. What is far from normal is the *inadequacy* of these processes with respect to threats that look unprecedented and overwhelming in size, urgency, and lethality. If there is at least some truth, and there seems to be enough of it, in the future scenarios of climate change propelled by global warming, then the present political process dealing with this phenomenon as it were business as usual reveals three major failures.

4. FAILURES OF MODERN POLITICS

The first failure is *cultural*: even more than with nuclear weapons, political actors around the world do not seem to have the broad-mindedness that is necessary to grasp the point that decision-making today has to do with a time prospect and with addressees, our fellow citizens of the far future, unknown to Athenian, Florentine, and later imperial policy makers of past history. This is what Günther Anders used to call the "Promethean gap" between the atom bomb, the new performance of man as Prometheus, and our still traditional and inadequate world of emotions and images of life (Anders 1956, 270, § 12). On a colloquial level, it translates into "lack of imagination," particularly into the lacking ability to imagine future times and problems to come. This has also to do with a way of policy-making that is still incapable to absorb science and scientific, even if divergent predictions as constitutive elements, not as occasional pieces of knowledge that can be stretched according to ideological or policy preferences.

The second failure is known as the *tragedy of the commons*,[38] a notion that can be perhaps reformulated for our purposes in these terms: the depletion of common goods such as the atmosphere, the protective ozone layer, and a relatively stable temperature is not seriously dealt with by the community of present users/spoilers either because the many particular interests of those who do not want to pay for the necessary changes prevail over the general interest in protecting those goods, or because the majority of present users/spoilers does not regard the interest of future generations as worth being protected. Particularly in the first case an aggravating factor lies in the mostly notional character of that community, which is rather a society of states, some of them being under certain circumstances more prone to tackle the problem as a shared responsibility than others, whose persisting notion or illusion of sovereignty makes them less permeable to accept and to share responsibility towards the remaining dwellers of the planet, alive or still to be born that they may be.[39] There is indeed a time dimension in the shifting of *sovereignty* from real to deceptive: the same states that today insist on their sovereignty (and, because of their present power, in a credible way) in order not to share responsibilities and burdens in reducing greenhouse emissions can in this way contribute to an earlier demise of their ability to protect their citizens of tomorrow (protection being the core performance of sovereignty), exactly because their actual and present sovereign decisions may prove wrong and bound to make the problem more acute. In the case of man-made climate change sovereignties are left with their traditional power even less than before the effects of nuclear weapons: it is as pointless to proclaim one's own neutrality in a large-scale nuclear war as it is to reassert that "the American lifestyle is not up for negotiation" is the appropriate approach to global environmental threats.[40]

The third failure could be consequently called a *tragedy of democracy*. Democracy is a successful way to regulate conflicts of principles and interests and to prevent them from becoming disruptive and bloody by allowing the majority to govern and the minority to have its fundamental rights respected (liberal democracy) and rerun the electoral game with a chance of becoming majority. A pre-condition for democracy to be regarded as the most just and advantageous method of conflict resolution is some stability of the social and natural environment: whatever majority can come out of the elections or the parliamentary vote, that environment shall not be significantly or irreversibly damaged or upset, and the next majority or coalition will have something (at worst, the damages of an unsuccessful war) to redress, but not to restart in a waste land. This stability is no longer a safe and implicit assumption (and that pre-condition, so far insufficiently highlighted by political theory because it was in fact obvious, needs to be worked out). Wrong decisions, significantly postponed decisions, as well as omissions (the refusal to put global environmental challenges on the agenda) can spoil the natural environment more than it is already, and contribute to cause irreversible damages. The same obviously applies to democratic decision-making that because of similar failures and drawbacks may result in unleashing or not preventing a nuclear war. In this light, democracy looks a less promising method of governance than it used to be in the pre-global challenges era and should be open to rethink itself under the pressure of the new challenges;[41] even if it remains more reassuring than other forms of governance because it allows for public debate on the agenda and pluralistic control of decision-making. All types of Leviathan are severely challenged by the new global threats in their basic protecting function, a fact which erodes their legitimacy. This is true for the democratic Leviathan as well, but is still performing better than the others, as proved by the circumstance that environmental protection and the related legislation have been invented and pushed on in the democratic countries with a free civil society. Yet, being better-performing than the others does not make by itself the level of protection provided by the democratic Leviathan adequate to the challenge.

Now, the new precariousness of that pre-condition raises a further problem to *traditional democracy*. A further presupposition for it to be the most legitimate method of governance was the coincidence of the decision-makers, the *demos* on election day, and those affected by the decisions, the totality of the citizens. Of course it was not just the actual citizens that were affected, but the next generations as well, at least one's own children and children's children; but the tacitly assumed stability of the environment put those affected in the present in the legitimate condition to speak for those affected in the future. What was democratically decided today used to be regarded as valid and good also for the citizens of tomorrow, while tyranny or oligarchy were considered to impose only the short-lived interests of the tyrant or the oligarchs. This is no longer the case, and it is correct to assume the possibility of a conflict of interest between generations: a fairly asymmetrical conflict, since the future generations are largely dependent on the effects of our decisions, while they have no instrument to coerce and deter us and do not even exist as present actors. They can exercise on us "moral deterrence," but it remains our business whether or not to act as their representatives in front of ourselves— a good reason to put that term between quotation marks.

If normal policy-making does not look sufficiently able to meet problems that are far from normal, this is not a good reason for outright proposing a new way of policy-making. Mere normativism, which wants to replace evil with good, or the belief that widely shared requirements can easily translate into policies can some time be useful to motivate political movements, but does not match the sensitivity of political philosophy for the problematic and deep-rooted nature of ultimate issues and difficulties. Its business is rather to explore what makes the very definition of what is good so difficult and perhaps irreparably belated, and what should change in our cognitive and normative images of the world in order to create premises that can be more favorable to positive change. This is why a philosophical exercise on climate change should not be expected to distill policy proposals, following an abstention-oriented line that will be further discussed in chapter seven. This means that this chapter does not tell us to choose agreements on targets such as emissions caps over those on actions, or a worldwide carbon tax over mandatory targets, as proposed by Cooper 2006a, or to approve or reject[42] emissions trading; not to speak of technical alternatives as between sequestration of carbon dioxide in oil wells or on the ocean floor *vs.* the "cooling of the planet" suggested by Crutzen (2006). On philosophical grounds only a policy choice can be excluded, that is the "only adaptation, no mitigation" thesis upheld by optimistic writers such as Lawson (2006) and Goklany (2007). The exclusion results first from the inability of any adaptation policy to make life on the planet still as decent as it used to be if CC2 with its upsetting physical transformations occurs, which is very likely if emissions are not seriously cut; then from the high improbability of adaptation aid from the rich countries being granted and effectively getting into the hands of those who need it; they are more likely to be abandoned to a somber destiny which we, supposedly well-equipped for adaptation, would have openly or hypocritically chosen to inflict upon them.[43]

From what we have so far discussed two questions have emerged that seem to have a decisive impact on whether or not we (the international community, humankind) should take action as to address in the appropriate scale the global environmental challenges. These questions are not related to a specific policy and rather require to go beyond the political science of climate change we have mainly discussed in this chapter and into the philosophical presuppositions of the actions required to meet those challenges.

The first question regards the possibility or obligation to take action under conditions of *relative uncertainty*, as the knowledge of climate change does not yet have the same level of certainty[44] attained in the knowledge of acid rain or ozone layer depletion. This question was discussed along with the related notion of risk at the outset of this book in chapter one: reconnecting what was said there with the specific uncertainty of climate change prediction and policy can be left to the reader. The second question concerns whether or not and in what measure we have obligations to future generations, a question which has also resulted from the nuclear weapons issue as crucial. It seems reasonable to discuss this fundamental normative question in Part Two of the book; but before we do this and close this chapter we should give our inquiry a broader frame, raising the philosophical question of what anthropogenic climate change may mean for the history of humankind; a question that is analogous to our previous attempt to elaborate on the meaning of nuclear weapons for modern political rationality.

5. SCIENCE, STATE, AND NATURE IN MODERNITY

Global challenges have represented a cleavage in the periodization of our evolution, as they signal the end of modernity or at least a major cleavage in it. To reflect upon this shift is itself a modern attitude, reflectivity being a defining feature of modernity since its very inception with the *querelle des anciens et des modernes*. What we are now facing is a twofold *querelle*: first, has modernity really come to an end, and should we therefore reshape the categories that helped us to tackle modern history, for example dismissing the realist belief in the centrality of the nation-state and looking at post-national and global patterns of coming evolution? The second debate regards the defining elements of post-modernity, which in this book turn out to be fairly different from the picture given in the two past decades by postmodernists.

To be true, we do not know much about the real evolution of that approach, as no comprehensively researched history of modern science, technology, and industry in their effects upon nature is yet available. We must be satisfied with a picture based on an early philosophical definition of the modern attitude, the *Baconian project*, as well as the present outcome of modernity, including most recently man-made climate change. We tend to match the original project with the recent physical outcome, but we do not indeed know how far the Baconian project, or rather the corresponding conventional wisdom has really influenced scientists, industrialists, generals, and statesmen along the centuries. We rather assume that they would have had a Baconian understanding of what they were effectively doing, should they have raised questions about reasons, goals, and values implied in what they were doing while pursuing a research theme rather than a different one, building a device or a weapon rather than other ones, or putting private stock or taxpayers' money in these enterprises rather than elsewhere. At least for the Western industrialism of the nineteenth and early twentieth century and the related positivist ideology of progress there are many clues that they indeed did so, particularly in the case of Fordism; but on the whole we should be aware that the history of the modern approach to nature, its study and its modification must have been more complex than it can be described resorting to the writing of a seventeenth century philosopher.

Several factors have been mentioned by innumerable authors as responsible for our present problems with the environment: science, technology, overpopulation, capitalism, the states system, and more recently gender patriarchalism.[45] How can they be put together in some kind of logical order of causation? A cultural evolution, the modern rise of scientific curiosity, came first and independently from the others, and only a bad use of sociology can look at it as just the consequence of an interest in dominating nature pushed ahead by the new commercial elites of Renaissance Europe. Copernicus, Galilei, and Newton were not executors of a Baconian conspiracy. Technological developments and industrial applications had a frequent interplay with scientific discovery, but it was interaction between independent variables rather than one-way causation. Technological developments came in the aftermath of discovery mostly so abundantly and so swiftly because the upcoming capitalist mode of production pushed innovation and expansion like no societal organization before, and some time between the nineteenth and the twentieth century the centralized and bureaucratic nation-state started putting more and more state funds in the fostering of scientific research and technological growth, but this did not make science a pure subordinate function of the capitalist enhancement of profit and labor productivity and/or the Westphalian state's preparedness for war. How does this all relate to man-made climate change?

As far as it goes back to human activity, global warming is the final result of the enormous increase of goods and services production since the Industrial Revolution and particularly the inception of mass production for the internal market in the twentieth century's social state and the opening of new markets abroad, lastly after the collapse of the Soviet empire and China's economic turn to capitalism. Overpopulation, which obviously multiplies by an expanding factor the emission of greenhouse gases, is not an independent cause, as it is rather the result of improved food, health, and transportation technologies, plus the effects of welfare policies in Western countries and international aid in some developing countries. Science, technology, and state all play a role in this increase in the transformation of nature and its unintended consequences, as they have so far shared three normative presuppositions:

1. The earth is an *unlimited* resource, both as dispenser of raw materials and container of waste.
2. Because economic and military power is their paramount objective, states are not poised to control *technologies* that may give them a greater power, except they pose an immediate and clear danger to their own present population or armed forces.[46]
3. Negative trans-boundary *externalities*, i.e. the damaging side effects of technologies beyond one's own borders, are not a concern of the individual state.

The environmental crisis of the last three or four decades has proven the first presupposition wrong as a matter of fact, and the third one unsustainable because of the conflicts (acid rain, sea pollution, and even more heavily global warming) that it has unleashed, while public control on technology has become either an achievement[47] or a matter of political debate and conflict, both nationally and internationally. At a deeper level, science and technology are no longer deemed to be morally neutral or by default laudable as they enhance knowledge and well-being: even if they are not by themselves evil or treacherous as some Green moralists may think, they are now not exempted from ethical questioning. Resorting to Max Weber's terminology, they cannot be approached just out of an *ethics of intention* (to do good to humanity or individual countries), as after Hiroshima and Chernobyl an *ethics of responsibility* has proven to be necessary in order to take care of their unintended consequences. Even if we do not give leeway to an overall refusal of science and technology and remain aware of how much they have contributed to the improvement of the human condition, gone is by far the credibility of the Enlightenment's project to transform the state of nature (both as external nature and enmity among humans) into a civil society based on science providing knowledge, technology providing goods, and the Leviathan providing protection.[48] Horkheimer and Adorno's (1947) word of the dialectics of Enlightenment turning domination over nature (*Naturbeherrschung*) into dependency on it (*Naturverfallenheit*) appears to have ironically expanded its original meaning into the future, as the unrestrained modern *Naturbeherrschung* has made us and, much more heavily, those who will come after us victims of huge and unfavorable changes in our physical environment. This means that if we do not want to follow the postmodern path and give up the importance of science, technology, and political institutions for shaping an acceptable life condition for humankind, we have to redesign the framework[49] in which they are supposed to cooperate and to acquire moral and political legitimacy.[50]

NOTES

1. More in general, climate is "a boundary value problem—a statistical description of the mean state and variability of a system, not an individual path through phase space." Schmidt 2007.

2. On the impact of climate on human life throughout the millennia see Lamb 1995.

3. IPCC 2001, 5–7. For a skeptical evaluation of the link between human activities and climate change see among others Burroughs 2001 and Leroux 2002.

4. In this case particularly about the amount of change compared with other similar events in earth's history and the causal link between increase of greenhouse gases and warming, a link that may come up in both directions. On the role of epistemic communities in environmental governance see Haas 1997.

5. Even less am I going to enter a discussion of Michael Crichton's *State of Fear* (2004), not so much for the reason that this indigestible conspiracy fiction seems to be a low point in Crichton's literary career, and rather because my book has nothing to do with cheap catastrophism and other vices of the environmentalist establishment targeted by the novelist.

6. This exceeded the natural range (180 to 300) over the last 650,000 years as determined from ice cones.

7. The extreme heat of summer 2003 in Europe had a death toll of about thirty-five thousand (Stern Review 2006, viii) aging or sick people, also because of the lack of preparedness on the side of the national wealth organizations.

8. For this concept see Dyson 2003.

9. In the vocabulary of climate science, mitigation means addressing the very causes of global warming, while adaptation refers to all measures aimed at making its consequences less damaging for humans.

10. Reference is to President George H. Bush's statement "the American lifestyle is not up for negotiation," released in 1992 in connection with the Rio de Janeiro environmental summit, as he was confronted with the proposal that the United States curb its huge energy consumption to reduce greenhouse emissions (quoted in Singer 2002, 2).

11. See IPCC 2001, 15–16, and cf. Burroughs 2001, 267.

12. Burroughs 2001, 260, makes future climate trend predictions on coupled atmosphere-ocean models, which would tell us more about oceanic thermohaline circulation.

13. As far as the European countries are concerned, the Kyoto targets are not too far from what they would in any case achieve by their own existing virtuous policies.

14. Even by maximum conformity to the Kyoto agreement the expected warming in 2050 would decrease from 1.4° to 1.395° (Jamieson 1992, 304). In fact, the countries that in 2003 were ready to ratify and implement the agreement accrued to just 19% of worldwide emissions, which makes their contribution to mitigation a symbolic rather than effective policy. For sure, symbolic policies can contribute to keep the chances for effective future action open, but their effectiveness should not be overstated.

15. A still little researched and debated case of interaction is that between global warming and global dimming, as the observed (since the 1950s) reduction of the amount of sunlight reaching the earth surface and caused by polluted air is called. Global dimming is supposed to be responsible for the deadly Sahel droughts of the 1970s and 1980s, but on the other hand it protects us to an extent from global warming, so that air pollution cannot be worked against if the latter is not at the same time mitigated, because otherwise clean air and more sunlight would push up the warming, see BBC 2006.

16. Oldfield 2005 takes the difficulties of studying and predicting climate change seriously, but also unveils the scientific flaws or easy tricks employed by some skeptics. In the preparation of IPCC 4 the methodological aspects of uncertainty have been carefully debated, see Manning and Petit 2003.

17. As free riding has been also investigated the intergenerational relationship between polluters and polluted, cf. Gosseries 2003.

18. Both aspects are worked out in Barrett 2003.

19. Unlike in International Relations, conflict means in this book politically organized competition on scarce resources, not just armed clashes and interstate wars.

This definition goes back as far as to Max Weber's notion of conflict (*Kampf*) in *Economy and Society* (Weber 1922, 38–40).

20. See Lamb 1995.

21. The following classification draws upon Homer-Dixon 1999 and Soroos 1997, who also provide a detailed phenomenology of conflicts.

22. This is not to deny that learning processes and policy design related to global conflicts can interact with those resulting from the local ones, as argued by Young 2002, chapter 6.

23. It also remains to be seen if changing temperatures on the Antarctica, caused by enhanced concentrations of greenhouse gases, is not depleting ozone in amounts that offset the positive effects of the Montreal regime.

24. For example Keohane et al. 1993 and Young 2002.

25. Cooperation under neither hegemony nor anarchy, it should be added, is cooperation beyond the neorealist horizon.

26. A useful account is given by Paterson 1996, chapters 5 and 6.

27. On the lacking enforcement of international environmental regimes cf. Schelling 2002 and Cooper 2006.

28. More on the ozone protection regime is to find in Parson 1993 and 2003.

29. "The monetary benefits of preventing future deaths from skin cancer far outweighed costs of CFC controls," as Barrett 2003, 230, put it.

30. This is the obsolete vocabulary still prevailing in political and even scholarly language, especially in the USA: as were states or international actors (how they should be mentioned in a correct wording) still identical with nations, as were in environmental policy the European Union, the major actor along with the USA, a nation or super-nation.

31. Some evidence of cultural differences in the face of threats, which are reflected in the divergent approach to global warming of US and EU: Europeans fear to be personally affected by global warming in the next ten years more intensely than Americans (73% vs. 64%); this is the only danger out of seven that is more feared on this side of the Atlantic, see Transatlantic Trends 2005, a thinner difference however against the divide between Democrats and Republicans in the US on the same question (73% vs. 48%). This picture finds correspondence in the Pew Research Center Survey on Global Warming 2007, conducted in January 2007 in the US, while interesting insights on how the concern for global warming correlates (however not directly) with the level of development can be found in the poll conducted by The Chicago Council on Global Affairs 2007.

32. For example, it is true that the shift from fossil fuel (gasoline) to hydrogen in car engines has been hampered or slowed down by oil-friendly governments and corporations, but as a matter of fact hydrogen-based engines are not only still unsatisfactory in economic terms, they also require energy consumption (from which source?) to produce hydrogen, and the greenhouse gases they release have to be somehow sequestered.

33. It took twelve years to Montreal from the first studies on ozone layer depletion by Rowland and Molina in 1975, but only two from the experimental evidence of the Antarctic hole in the ozone layer discovered by Farman in 1985, see Barrett 2003, chapter 8.

34. Cf. Weart 2003, 152–54.

35. This seems to be the case with the Katrina hurricane, fully, if wrongly attributed to global warming in some corners of the American and worldwide public opinion. One more evidence of all things human being so entangled that from evil (the tragedy of New Orleans, the hurried indictment of global warming for it) good (more openness toward the problem of climate change) can be born.

36. Young 2002, 156.

37. For example raising a carbon tax so heavy for the consumers that it forces them to burn less fuel by improving the related technologies.

38. This notion was first introduced by Garrett Hardin's famous namesake article of 1974, cf. Introduction, n. 8. The problem of unregulated (by property rights or government authority) access to global commons as source of environmental change is discussed in Paterson 2000, 23–26.

39. Soroos 1997 denotes as "environmental tragedy" the situation in which multiple actors are permitted to derive unlimited private gains from exploiting a domain having finite resources.

40. On neutrality cf. chapter three, § 3.

41. This is the reason that makes recent (e.g. deliberative) theories of democracy unsatisfactory: they think of it as just a relationship between citizens, groups, and institutions, with no reference to the changing substantive problems democratic power has to come to terms or struggles with.

42. As suggested by Schelling (2002), whose opinion this author finds convincing.

43. More on this point in chapter five, § 3, in the remarks on vulnerability. Adaptation not disjointed from fairness is unmanageable also because of the increase of the world population to 9.1 billion by 2050, with the additional 2.6 billion all living in the poor countries, see Cohen 2006.

44. Needless to say that in the post-Kuhnian era "scientific certainty" has lost the objectivistic or positivist connotations it might have implied before, but this is here irrelevant to our point.

45. See Paterson 2000, 40–54.

46. At the basis of the Geneva Convention of 1925 that restrained gas warfare there was the awareness that this chemical weapon, easily available to everybody, was self-defeating as it did not give any true competitive advantage, see chapter three, § 5.

47. Ministries of the Environment or equivalent institutions like the U.S. Environmental Protection Agency have been created in most if not all countries and issuance of industrial and agricultural safety regulations is a major activity of the European Union, in postmodern times the most advanced experiment in governance with little government.

48. This is the Baconian project in a secularized standard version, that is, minus the religious perspective of recovering Eden on earth as it was designed by the very Francis Bacon, cf. Merchant 2003, part 1, and Perez-Ramos 1996, 327–30.

49. On the epistemological problem of how at the end of modernity science can adapt to a world in constant flux, far away from that of classical academic sciences, without giving itself up, see Lelas 2000, chapter 15.

50. I will pick up this discussion again in chapter six. § 3.

Part II

PHILOSOPHICAL AND POLITICAL PERSPECTIVES

There is little that can be said in advance as an introduction to this part. Unlike the first one, it does not elaborate on existing knowledge (history, International Relations, climatology) in order to bring about an interpretation of the world shaped by global challenges; it is rather dedicated to elucidate not only the ethical, but more generally the philosophical dilemmas created by them. How the argument I am trying to construct and find good reasons for unfolds cannot be summarized but only followed in its complex development. This is valid to an even larger measure for the unconventional kind of relationship I am trying to delineate between philosophical and ethical conclusions and political suggestions.

As to the distribution of the subject matter, chapter five is an extension of the "Hobbesian moment" of chapter two in the field of normative ethics and makes the case for seeing in the survival of humankind rather than in the establishment of justice the primary normative goal. Chapter six goes beyond normativism and highlights the meaning (in common as well as in philosophical sense) that working for that survival under civilized conditions might have for the identity of present and future human beings. Chapter seven finally gathers up the threads of all political problems and solutions hinted at in the previous chapters, however ending with a warning to the reader to expect neither messages of doom or salvation nor ready-to-use projects from political philosophy.

Chapter Five

We, the Future
Generations, and Humankind

Future generations as an addressee of our concern and sense of responsibility
or even obligation have come up in this book in the different contexts of the
two global challenges examined in the previous chapters, and so has the admittedly vague notion of humankind. Nuclear weapons and man-made climate change have fundamentally changed the real state of affairs in our attitude to future generations, as for partly similar, partly different reasons they
both let emerge the possibility that we pre-define the very environmental or
civilizational fundaments of their life in a way that would make our own life
impossible or unacceptable to ourselves. What I am hinting at is a world in
which civilization has been widely cancelled by nuclear war and nuclear winter, or is going to be permanently unsettled by catastrophic consequences of
man-made climate change. It is of the greatest importance for the correct understanding of this entire part of the book to hold firm that by civilization I
do not mean any order of the human community based on particular values
and principles, or implying a sense of progress towards the fulfillment of the
essence of man (and woman). As already mentioned in the Introduction, by
civilization I simply mean the fact that, unlike other animals, human beings
cannot survive without a certain degree of mental development, language-
and tool-based control on their life environment, and social communication
and cooperation among themselves. I am keeping any moral or metaphysical
value assessment analytically separated from this basic anthropological element.

We are now taking a closer look at those two philosophical (but also legal)
concepts, obligation and responsibility, asking what they mean to our approach to global challenges as well as how far these challenges may induce
philosophy to revise some of its own settings. After summarizing the philosophical discussion on future generations (§ 1) and discussing the current

normative approaches to this issue (§ 2), I shall seek another foundation for our obligation towards our posterity and argue that global challenges should also induce various theories of morality to recognize a new imperative or rather meta-imperative telling us to do our best for humankind's survival (§ 3). I shall finally argue that the two global threats give a new, not-notional existence to 'humankind' (§ 4). Yet it is not only moral (normative) philosophy that can provide us with guidance as to what to do in the face of global threats. Equally important are questions (see chapter six) regarding the meaning of our individual and intergenerational life projects as well as our identity as a global group.

1. HOW GLOBAL CHALLENGES
CAN AFFECT FUTURE GENERATIONS

Ever since John Rawls' *A Theory of Justice* was first published in 1971 with two paragraphs[1] examining the problem of justice between generations, this topic gained full citizenship in philosophy and has been discussed either at a high level of philosophical abstraction, principally in Derek Parfit's *Reasons and Persons* of 1984,[2] or as a matter of applied ethics, that is, with regard to the problems of food, energy, and more recently environmental protection. A preliminary remark: this chapter is not going to be a further step in the main line of this debate nor its "application" to the topic of global challenges. On the contrary, the very nature of this matter requires reconfiguring it along its specific lines. The first reason for this shift is that, unlike most of the literature we are alluding to, we are not dealing here with problems of distributive justice, if distribution refers to goods produced by the present or past generations, which we can decide to distribute between us and the future ones according to a just or unjust *ratio*. The burdens that may derive from attempts to eliminate the threat to ignite a nuclear war or to divert the Gulf Stream cannot be "justly" distributed between our posterity and us. In both cases indeed to imagine that dealing with threats to our survival can be postponed on behalf of a balanced intergenerational burden-sharing is pointless and self-defeating, as the threatened event can occur in full strength while we wait or even because we have waited to act; moreover, in the case of global warming we have learnt that it may be too late even if we would seriously start reducing emissions today. A further reason for not simply joining the mainstream debate on future generations is that in my reasoning the normative approach will turn out not to be the exhaustive approach to the problems we may have with them. This is not to dismiss the legitimacy of any research on intergenerational justice, but to point out that: a. global challenges allow for a very specific set of problems, which are not a sub-chapter of that research; b. one thing is to raise the problem of justice between generations as a necessary development to the theory of how to build a society in which freedom, wealth, and welfare are justly distributed; another thing is to look at two threats to the physical survival of our civilization, if not our kind, and to investigate the normative questions they raise.

There are further aspects of that ethical debate, such as the "discount" and the "identity of future generations" issues, in which the direction proposed in this book will go beyond the usual terms of that debate; we shall go back to them later on. It is perhaps useful at this point to highlight the concrete differences between the areas of justice in the general debate on future generations and what is at stake for them when it comes to global challenges. In the first case it is the welfare benefits deriving from the accumulation of wealth in human societies over time or a more rational and respectful consumption of natural resources such as fossil energy sources which are not renewable, but can be replaced by technological innovation, for example, the hydrogen motor or, as some believe, "cold fusion." What is at stake with global challenges are in the second case what we may call "survival goods," most of which are to be defined in negative terms: to remain alive, not to be born with genetic deformities, not to be exposed to lethal pollutants and oncogenic radiation, not to be deprived of a vital amount of water, not to be subject to continuous and ruinous temperature change causing large flooding and spreading of diseases. Furthermore, as human beings are essentially cultural animals and experience their lives in a collective and communicative way, their physical survival makes sense only if it is not accompanied by a permanent and unrestricted state of fear, if it does not occur in a "state of nature" dominated by the perspective of nuclear genocide or the expectation of deadly cold/heat.

We are now going to take a closer look at the specific and somehow different features of the future generations problem in the two global challenges.

In front of the threat of nuclear war our problem with future generations is not really whether or not we allow them to be born,[3] thanks to the circumstance that we abstain from burning the planet before they have the chance to, but rather if we want to do our best to free them and us from a risk that affects both, though it may in this moment endanger them rather than us. At first (we shall soon qualify this remark) it looks like it does not take any altruistic sense of obligation to do something that is suggested to us by our own sense of self-preservation, in a case in which we have no costs to bear that are not paid back in terms of our security and we need no particular reason to do for them something that we are going to do anyway for ourselves. Admittedly, all this can be contested by two very different critics: those who believe that nuclear deterrence or even proliferation is a safer and more effective security system than anyone else, present or future, will disagree with eliminating nuclear weapons either physically, by full disarmament, or politically, which last could mean to confer them to some kind of supranational authority. This belief is perhaps not so predominant now, nearly twenty years after the end of the Cold War, but played a role among realist International Relations scholars immediately after 1989.[4] Disagreement is to be expected also from those who see no good reason for preserving humankind from extinction, may they be nihilists in the wake of the Marquis de Sade[5] or "deep ecologists" wishful of making the planet free from the oppressive and polluting humankind. The first kind of opposition to the anti-nuclear obligation[6] was for a long while a politically relevant argument, but is now weaker, in the sense that at least among the five "official" nuclear countries and Israel nuclear weapons are seen as a sad necessity, not as a first choice weapon.[7] The nihilist argument is and will probably remain politically pointless and find no supporters even among present-day terrorists, but has a philosophical relevance which will be discussed later in this chapter.[8]

Things look quite different with man-made climate change, that is, with the obligation to avoid and prevent depletion and decay of features of our physical environment that are essential to the life of all mankind. Discounting the implementation of this obligation over time makes no sense for several reasons: first, we have seen in the previous chapter that it may be too late to be effective if we would start right now curbing emissions, and doing so gradually across decades or centuries would be definitively ineffective. Secondly, if those features are life-essential, the "last generation" has to be endowed with them in the same amount as the first one, our generation; here is no risk of "chronological unfairness," quite the contrary holds true.[9] Finally, unlike local environments, which can selectively be polluted in a trade-off for more growth and less poverty for the next generations, the global environment cannot, because greenhouse gases emitted by Bangladesh or Burkina Faso will add to harmful climate change there as well as worldwide, now as well as in a hundred years.

As to *costs and benefits* calculated from the venture point of the present and the very next generations' self-interest, relief from the curbing of emissions is rather local (e.g. for highly polluted urban areas) than global, as the effect on global warming will be delayed by the physics of climate and its benefits shall first accrue to people supposed to live in a pretty far future. A boost will be given to environmental protection, including the related technologies and industries; some of the losses in GDP that the Stern Review, as we have seen in chapter four, sees as a consequence of global warming, will be neutralized. But on the whole this seems so far not to match the losses in GDP and growth rate forecast for example in case of full implementation of the Kyoto agreement; this will include a major industrial restructuring with lay-offs and a heavier impact on developing countries.

Unlike in the case of nuclear weapons, steps which will come to the advantage of future generations cannot be justified on grounds of our self-interest, as a beneficial side effect. Either we redefine our self-interest in more idealistic terms or we have to go outside of its sphere and find well-carved justifications for taking those steps. From a *motivational* viewpoint there are further elements which make the case for tackling the nuclear issue stronger: first, we know more, with higher certainty and less controversy[10] about the effects of nuclear war than about global warming and its effects. Then, nuclear war is a non-gradual event: it may not happen but we know that if it happens it will be full hell, at once and without respite. Global warming may be already on its way, but its consequences will hit us and the generations next to us only gradually and with important geographical differences; more devastating effects such as the freezing of the British Isles may come or not come and will in any case hit generations pretty far from us.

Now, before we look for possible justifications of obligations to future generations beyond our self-interest, let us qualify what we said about attempts to get rid of the nuclear threat being based on that interest. Qualifications are necessary as soon as we shift our attention from the normative obligation to its political implementation, which does not only mean containing proliferation, but rather pursuing a long-term strategy of eliminating the nuclear threat by dismantling the nuclear deterrence regime, and inducing all nuclear powers to eliminate their arsenals or to confer them to a supranational authority.[11] This can be delayed or *sine die* postponed or only halfheartedly pursued because that threat appears not to be severe enough for our generation, as it does after the end of the Cold War. Or putting the nuclear question high on the agenda may imply costs that a country's or alliance's political leadership regards as too high, such as giving up one's own nuclear power or entering a partnership with otherwise unwelcome countries. To take the step from the principle of liberating humanity from the nuclear threat to its realization appears to need additional justification in an obligation to future generations, whenever the justification based on self-interest does not provide enough motivation because other aspects or interpretations of the present generation's interest prevail.

At the end of this survey of the ways the two global challenges can affect future generations, our however enlightened self-interest turns out not to be a sufficient platform for justifying and motivating actions that may prevent harm to them. We shall now look for arguments that go beyond our generation's interest. Or at least beyond our interest in survival and welfare, that is, our selfish or—in everyday language—"material" interests, which mostly identify with interests altogether. We may as well have "ideal" interests, that is, be interested or have a stake in pursuing and realizing something that is important to us, but does not imply any advantage for us. Making the world safe and livable for the far posterity is such an ideal interest.

2. NORMATIVE APPROACHES

At the beginning of § 1 some substantive differences between our investigation and the general debate on intergenerational justice have been outlined. On the ground that we are now concerned about warranting survival rather than doing justice over time, we can now state some basic rules for our search after the best argument in favor of future generations. First, it must make as little initial assumptions about them as possible, otherwise it will

a. build not enough *consensus* among those (contemporary or future debaters) who do not share our philosophy of man, history, and society;
b. be with more *likelihood* at risk to collapse as soon as new developments or scientific discoveries disprove one or more of those assumptions.

These are special reasons that converge with the usual motifs of argumentative economy and elegance in advising us not to inflate our premises.

A second rule takes note of the circumstance that in this ethics debate factual assumptions about future events are intertwined with normative statements, for example, the hypothesis of humankind coming to an end, thus eliminating the problem of future generations, because in a particular ethical doctrine human beings are assumed to be free not to reproduce themselves. When it comes to factual assumptions like this, I shall follow the rule of checking them against their likelihood, be this based on empirical evidence or structural constraints. In the case mentioned I shall dismiss that hypothesis because of both reasons: it is as a matter-of-fact impossible to keep all and every man and woman on the planet from generating and there will always be at least some human beings on earth or elsewhere, at last until the solar system collapses. Their existence, even in a very limited number, is morally as relevant as to give ground to our concern for them. Moreover, even if in abstract moral reasoning the possibility of a zero degree in human reproduction can be granted, from the political point of view no procedure is imaginable under which a decision in this sense can be made and implemented, let alone legitimized—except perhaps under a perfect and seamless totalitarian world government which happily we can for the time being leave to (political) science fiction to imagine, as present developments and predictions point to an anarchical globalization rather than planetary scientific dictatorship. There is good epistemological sense in this checking of moral assumptions as soon as they make assertions regarding empirically verifiable events or other mental environments, such as politics and political philosophy. Last not least, this is a way to get rid of the *futility* that sprouts in some corners of pure, unchecked normativism.

There is finally a third rule: the more theories justifying obligations allow for arbitrarily chosen pre-conditions, the less they fit the need to justify obligations to future generations with arguments that are stable and reasonable enough as to be translated into a *public argument*, capable of supporting a political strategy. For example, in Rawls' theory of justice and among his followers uncertainty prevails as to whether the parties in the original position are to be regarded as *contemporary,* as Rawls put it,[12] or rather under a veil of ignorance with respect to their being or not contemporary. This second configuration of the original position is suggested by those[13] who wish to implant the concern for future generations in the initial settings of the theory of justice. That a theory can have different normative outputs according to the alternative inputs among which we can choose may be normal for post-metaphysical philosophy, but it does not make that theory particularly appropriate to issue unequivocal indications of obligation that may translate into policies. Those who should endorse those obligations and policies will see in the different outcomes the influence of various philosophical or ideological parties rather than a neutral response to threats and questions concerning everybody, and sense arbitrariness rather than universal validity. In any case, the choice between different initial settings turns out not to be a matter falling inside normative theory itself, but rather something depending on the cultural environment the partners to the theory come from, or on the image of man they may agree to share, this last source being exactly what we shall discuss in the second part of this chapter. If this holds true, pivotal for the obligation to future generations issue appear to be in the first place — as we shall see in the next sections — the cultural values and anthropological images we agree upon and in the second one the normative arguments we may then develop within these frames of reference.

This all leads us to better understand a deeper concern in the moral consideration of global challenges and a more general rule, which underlies the three we have just examined. Our concern is to find reasons justifying obligations to future generations that are not peculiar to particular theories of morality as well as their substantive values and goals (such as distributive justice, freedom/autonomy, the common good) and do not have to share their lot (to be accepted or refused along social or ideological party lines and later altogether dismissed or forgotten). The reasons we are looking for are sober in acknowledging their own presuppositions (an anthropology slimmed to essentials), minimalist in formulating their goal (the survival of present and future generations, whatever moral and political regime they may want to choose), as color-blind as possible (as to gain the largest transcultural acceptance ever possible); in one word, *thin* rather than *thick*, yet as thin as robust. This is not simply because the more and more time-enduring acceptance you want, the less partisan your views have to be: actually, if this translates into watering them down, the reasons you choose on this ground may turn out to lack substance. It is rather that the more existential for everybody's survival the challenges called global are, the less dependent on more advanced goals and values (how to shape a just society, how to enhance solidarity or brotherhood among human beings, how to approach all sentient beings with equal respect) your arguments and norms have to be. Moreover, the two global challenges, though they derive from human activities, result in physical threats that are empirically verifiable and measurable; their genesis can and will still in this chapter be interpreted in historical and philosophical terms, but to recognize them as threats that must be neutralized is different from, say, the attitude that condemns this society because of its moral evils (egoism, narcissism, alienation, intolerance for the Other, cultural imperialism and the like) focused upon in our different moral and social conceptions.

This book proceeds the way we have just described because it examines the lot of modern humanity at its end station starting from the more or less severe, but predictable and widely recognized challenges to its existence under acceptable conditions rather than setting up a systematic evaluation of its evolution and the values we may want to promote in it. This is not to promote the illusion of just designing some kind of theory-independent and value-free problem-solving. Whenever necessary, I shall declare which moral postulates (respect for humankind, present and future) and theoretical frame of reference (my own version of the theory of a political modernity, which has been already hinted at in chapter three) underlie my formulations of emerging problems and possible courses of action. However, especially in normative issues, keeping one's own postulates and values on the thin rather than the thick side makes a difference.

Having clarified certain standards that I am going to use, I shall now check how various theories of moral obligation are or are not able to justify the obligation to future generations that has emerged as a possible or even necessary choice from the assessment of the impending global threats.

In the light of those standards it is not difficult to see that *teleological theories* of the "good life" are the least helpful in our case. Teleological and particularly arethaic theories are in general better (if any) fit for everyday life rather than for ultimate emergencies such as nuclear war and global poisoning of the environment, which rather seem to require focusing on the rightness of actions, as action-centered theories of what is just to do by definition.[14] In any case, theories of Aristotelian inspiration are for our purpose too thick in presuppositions and goals, and so are eminently moral theories based on religious faith. Moreover, making our present care somehow depending on how far we can include future generations in our plans for achieving the good life infringes the minimalist universalism which our moral common sense makes almost everybody share: an elementary sense of respect for human beings just because they will dwell on the earth tells us not to leave it in unliveable conditions, whatever their preferences, ideas, and conceptions of the good may be, and also whatever fate our plans for a good life may have. These are also the reasons why the communitarian way to justifying our obligations to future generations put forward by Avner de Shalit in his otherwise useful book does not meet the standards raised above, particularly when he outlines a transgenerational community, based on moral and cultural similarity and binding together us and our posterity, as the firm ground on which our obligations should stand. Shalit is only coherent with his approach as with regard to generations of a very far future, whose similarity with us can hardly be assumed, he sees a weaker obligation of humanity replacing the stronger obligation of justice we have towards our "communitarian" posterity.[15] Before any moral consideration, this different treatment is physically pointless, because the prevention of major damage from global warming to people assumed to live in a thousand years must begin now and be steadily pursued if it has to reach any effect and fulfill the meaning we put in taking economic sacrifices upon us in order to bring down greenhouse emissions.

If we turn to the normative, action-centered approach, we are first confronted with deontological models of justification. The original Kantian categorical imperative is as such no longer available to us, because Kant's robust and substantive idea of Reason has not withstood the shift towards a weaker concept or the wholesome renunciation of it. Even if one takes a slimmed-down version of Kant's approach, which retains just the requirement that maxims of action must be universalizable (the so-called U-test), a Kantian deontological ethic would need to be rewritten introducing in it the two dimensions of time and risk that must lack in its original version, as they were not actual, not even imaginable in Kant's lifetime: the being-at-risk of present and especially future humankind, the origin of this risk in perverse, unintended consequences of human action. This is only an announcement of issues we shall soon turn our attention to and shall not hinder us to pick up Kantian notions that may help us to clarify further steps in our inquiry. On Kantian grounds remain at least partially the two leading deontological theories in present moral and political philosophy: the ethics of discourse designed by Jürgen Habermas and Karl-Otto Apel, and John Rawls' contractarian theory. The first one is by its very interest (to find the foundation of ethics in the inner structure of the communication) not at all involved in questions of (counter-intentional) consequences of action, responsibility, and future[16] and seems under-equipped to come to terms with substantive and time problems of collective action.

Contractarian theory, as mentioned at the outset of this chapter, has put very early the future generations problem on the agenda, but mainly in the version bent on the just savings we have to set apart for them. The problem can be addressed in two different versions: First, as a contract between generations—as if we could reframe as a contract Burke's idea of the partnership between past, present, and future generations that only makes possible to realize the partnership's ends;[17] but this can be at best intended metaphorically[18] and is indeed excluded by a series of factors. The partners to the original position are individuals, not generations, and the veil of ignorance covers just the generation they belong to, but not the circumstance that they belong to the same generation. The partners are "free and rational persons concerned to further their own interests"[19] and to pursue mutual advantage, which must be by definition ruled out in the case of future partners, as there cannot be any reciprocity between them and us; they cannot do any harm or any good to us, except we turn our attention to ideal goods such as good or bad fame, which posterity can attach to its memory of us. But we can just hope or fear this, since a binding agreement with yet non-existing persons is unthinkable. Also, in the case of humankind the further circumstance put forward by Rawls to explain our relationship to future generations, that is, that "the life of a people is conceived as a scheme of cooperation spread out in historical times,"[20] does not apply, as humankind, being fundamentally divided into peoples, cannot be conceived of as one people, nor its life, which can end in catastrophe, as a scheme of cooperation; cooperation among and political integration into humankind have to be created and fought for, being by no means intrinsic to the notion of humankind.

Rawls himself saw the obligation to future generations not as the outcome of a contract with them, not even of an obligation to our immediate descendants,[21] but as a qualified consequence of the social contract among contemporaries. The very prospect of a just society requires that "each generation must not only preserve the gains of culture and civilization, and maintain intact those just institutions that have been established, but it must also put aside [by applying the difference principle over time, F.C.] in each period of time a suitable amount of real capital accumulation" (Rawls 1971, 285).[22]

The obligations to future generations, which in *A Theory of Justice* are defined with regard to our just savings in their favor, can perhaps be easily regarded as if they would include giving them a chance to be born as well as preserving the earth in a hospitable state, that is,[23] "still enough and as good left"—even if Rawls never makes this explicit, which makes this interpretation conjectural and also raises the problem of the meta-imperative we shall address later on. More extended benefits would accrue to them if, as proposed by the critics of Rawls mentioned above, we change the settings of the original position, that is, if we conceive of the parties as non-contemporary. Within the framework of Rawls' own version of contractarianism those obligations can be better argued than within the metaphor of an intergenerational contract, but to do so we need first to "buy" Rawls' entire approach, be it in the original shape of 1971 or in the later one of the nineties.[24] This makes contractarianism too "thick" an approach for our needs (to find the least theory-dependent justification for what we owe to future generations) and purposes (to ensure that respect for future humankind does not hinge on ideological, religious, or however particularistic and substantive premises). Rawlsian contractarianism is more open to these needs and purposes than the ethics of discourse, as we shall still see, but it failed to thematize them because the historical settings of these two post-Kantian theories remain centered around the problem of social distributive justice in the domestic dimension or, in the case of Habermas and Apel, the establishment of an advance defense (the idea of communicative action) against authoritarianism and/or moral relativism; in neither case do the unwanted consequences of technical/instrumental or power-oriented action[25] become the leading interest in theory building. While the *social and historical ambience*, or in other words the set of problems to which a theory of justification is related as its often undeclared background, does not tell anything about its rightness, it is relevant to its usefulness in public discourse, and global challenges still have to stimulate the creation of a general theory of morals that focuses on the normative questions they raise. What we can find so far is either the still interesting debate on nuclear ethics in the eighties, which however did not go beyond the dimension of applied ethics,[26] or first and still rough attempts at shifting paradigm from anthropocentric to biocentric in ecological ethics.

Our dissatisfaction with normative theories does not recede in front of *utilitarianism*, an approach that is even less able than others to grasp the morally dramatic specificity of the global challenges. The postulate that makes utilitarianism interested in the greatest possible happiness of the individuals regardless of when and where they live is too generic on the one hand as to focus on the obligation to work for the survival of humankind, while it can on the other hand lead to the "repugnant conclusion" as famously defined and criticized by Parfit:[27] more people on the planet with "lives that are barely worth living" are better than a lesser number with a high standard of living because the sum of happiness is greater in the first than in the second case.[28] As if morality were the numeric maximization of an impersonal, disembodied "happiness." Secondly, the utilitarian imperative to maximize the well-being of future people tends to ignore one aspect of the so called "identity-problem," that is, the possibility that their preferences may be quite different from ours, thus making the content of our obligation, as far as it consists of the projection of our preferences into the future, either oppressive (we superimpose our opinion of good and evil, pleasant and unpleasant over posterity) or ineffective (we take over us sacrifices to ensure posterity's happiness but they turn out to be unhappy over what we have chosen to set apart for them) or both. This is a problem affecting whatever intention or obligation to act for the welfare of posterity we may heed, and whatever justification we may turn to. Even Rawls' cautious assumption that there are certain social primary goods, "that is things that every rational man is presumed to want [such as] rights and liberties, powers and opportunities, income and wealth [and self-respect]"[29] may result time-parochial, as the notion of what a rational man (or woman) is may well change over time, especially in future eras of rapid technological revolutions (let us only think of bioengineering) that are supposed to affect the very basics of human life as it has been so far. Our interest is indeed focused on the natural primary good of living under decent environmental conditions, a good which obviously can be attained only under certain conditions (of civilization) and makes sense to our posterity only if some essential cultural-anthropological features of human beings are preserved. Which features? In an essay which remains important more than a quarter century later Gregory Kavka advanced a list of the features that make human beings alike beyond all cultural diversity and will be shared by future people for "very many generations": "their vulnerability to physical and mental suffering and to death, their capacity for enjoyment (including the enjoyment of complex activities and interactions with others), their self-consciousness, their capacity for long-range purposive planning and action, and their capacity for cooperation and identification with others."[30] I suggest to conceive of these features as *anthropological basic equipment*,[31] which first makes human beings capable—as Kavka put it—of entering into moral relations with each other, where "moral" just means the ability of understanding and implementing rules of civilized coexistence. They are the least substantive and engaging assumption we need to make about future human beings when we think of assuming obligations towards them, if these obligations are to make sense and the price we may have to pay is to be justified. For sure we cannot pretend full certainty about our fellow humans preserving those features, so that in a sense we *bet* they will be like that (i.e. like us) and have no tools to exclude a future upsetting of that minimal anthropological equipment. But, having reduced the substantive assumptions about our posterity to a minimum, it seems to be a reasonable bet. On the other hand, without this minimal degree of anthropological similarity it would be difficult to justify any particular concern and obligation towards beings that we would have to regard as sentient, but not necessarily human in the cultural sense of this word.

It is on the background of this anthropological continuity that we have to look at the survival goods mentioned in § 1 of this chapter as the content of our eventual obligation to future generations. Against the scenario that results from this combination, the normative approaches we have examined appear to be at a time not only too thick in the meaning we have tried to clarify, but also too thin, as they justify the obligation (when they do so) in terms too little specific to the problem configuration proper to global challenges. We shall soon explore if we can find a more specific way, but first let us ask if those approaches have something to learn from their confrontation with these challenges. Normative doctrines of morality have, as we have seen, an uncertain or only implicit and unspecific relationship to the problem of future generations as defined by the global challenges; this is typical of moral (and more generally philosophical) doctrines in which traditionally hidden presuppositions have been put in doubt. The continuation of humankind is no longer an obvious matter-of-fact, even less the permanence of civilized conditions that first make morality possible. On a planet upset by nuclear winter, the normative philosopher who might still uphold his or her system of morality and justice would be deservedly received by the few survivors with words echoing Rousseau's curse against those who in their books praise civil institutions credited to bring peace and justice, but remain blind to the effective horrors of a battlefield.[32] Normative systems cannot ignore that an unexpected evolution in our real conditions of life, that is, the emergence of man-made lethal threats to humankind, has put in question among other things the legitimate continuation of their traditional abstinence from substantive problems and positions and prevailing orientation to procedural rules.[33] This is not equal to the triviality that without human beings who are alive and are not back in the stone age[34] no moral doctrine would find addressees; it is rather that the self-elimination of the human race and/or civilization would prove untrue the assumption that human beings are capable of acting morally and worth of being taught how and why to do so (in the doctrines of obligation and justification). As a conclusion, normative theories of morality seem now to have to introduce not as one of their possible outcomes, but rather as a preliminary imperative "Do your best to prevent the termination of human life and civilization by human hands." As this norm regards the condition that first makes morality possible, it seems appropriate to define it as a *meta-imperative,* that is, an imperative that comes before any other particular norm and whose absence would void all other norms and rules. It is not possible to analyze here if and how far the introduction of the meta-imperative set a restructuring of the normative theories we have briefly discussed in motion. On the one hand it must be underlined that a meta-imperative is still an imperative, that means that it cannot be handled as a harmless presupposition, after bowing to which moral theory can go back to business as usual, as it has done after the blasts on Hiroshima and Nagasaki.[35] On the other hand it is worth stressing that it is not the effective avoidance of global catastrophes, but rather the thematization of the problem, now that it is undeniable, and the response given to it by the meta-imperative that can reinforce the foundation and the credibility of moral theories. These are not as time-neutral as they pretend to be; Aristotle's *phronesis* was expression of a society and a human type different from those of Kant's categorical imperative, even if both have outlived their own time and still help us to penetrate our moral world. The postmodern scenario of global threats still requires a rewriting of the moral agenda in which they are taken seriously.

3. PATTERNS OF OBLIGATION

Restructuring the received normative approaches in order to accommodate the new global challenges is a road I am not going to further explore, because the results of the restructuring job may be disappointing, while we do not need to start thinking under the overload of thick premises and inputs. I am instead going to experiment with another path or net of paths, which is however not alternative to the main systematic road. Let us look at those effective and familiar patterns of obligation that are at work in our communal life regardless of where (and even whether) we locate them in our doctrinal systems, and unburdened by "identity problems." This last clause, as we have seen, has a twofold explanation: I regard the problem "if" there will be future generations as a false, even futile problem, inasmuch as the "if" is fictitiously made dependent on a collective act of will resulting from rational discussion. Secondly, the problem "what" they will look like is brought down to a minimal dimension inasmuch as we do not attribute to those people any particular preference, let alone project our own preferences onto them, but think of them as endowed with the basic anthropological equipment and seeking survival goods.

Trust, vulnerability, and responsibility are the three keywords to put to test in order to find out if our obligation to give the generations that will be born in the future essential life chances "still enough and as good" as those we found on the planet can be tied to existing and effective patterns. *Trust* is a fundamental informal element of society's connective tissue, a necessary complement of formal institutions such as law and government.[36] We are looking however at a particular, though seminal trust relationship, which is at a time biological and cultural: giving birth to children or approving of it (in case of grandchildren, nephews and nieces, or at the "colder" end of the spectrum upholding welfare legislation for the family), we give children ground enough to trust us and to expect from us a caring behavior.[37] This is not just a psychological inclination, but a legal institution as well (from Roman law to the United Nations Convention on the Rights of the Child); as a ground belief and a social practice it is shared across cultures worldwide and the departures from it (infanticide parents and relatives, murderous pedophiles, child traffickers) do not undermine the universality of this trust relationship.[38] Trust is also meant here as a unidirectional relation: we give our children a ground and a right to trust us, to expect from us that we give them life chances on the planet as good as ours or at least decent, and they will probably repay us with some affection and care (social care, e.g. pensions, or in the United States Medicaid, and not just psychological). But their obligation to us is of a much weaker nature than ours to them, as they have not asked to be born, whereas we are the only cause of their coming into the world. This form of trust is nonreciprocal and in any case strongly asymmetrical. The only thing we can count upon is that they will reproduce with their own children the same trust-generating relation we have with them, our parents with us and our ancestors with their children. It is important to understand that this expectation is not based on our confidence to be able to convince future generations to endorse one or several of our "thick" theories of morality, obligation, and justice, in order to provide their posterity with survival goods; we do not need to enter this bet—and to lose it, as it is risky to make the life chances of future men and women dependent upon our changing philosophies and their lot.

   Now, on the ground of this pattern of obligation deriving from the trust every mother and/or father infuses into their children in any generation, we can first assume that as a matter of fact concern and care for immediate descendants are a moral constant across generations, not because they constitute all together a moral community, but simply because trust and obligation are reproducible and in fact reproduced patterns along the intergenerational chain.[39] We may see our posterity in five hundred years as fully stranger to us, but we know with enough approximation that, if humankind will have not ceased to exist by then, those beings[40] will relate to their offspring in a way not substantially different from ours. This chain represents, however, just a credible assumption about a morally relevant fact, which we derive analytically from our own relationship to the next generation; so far no new and specifically normative or deontic element has been added. The chain is enriched and reinforced by introducing a second element, *vulnerability*.[41] This moment, a development of the anthropologic basic equipment mentioned before, is essential to whatever obligation we enter towards people (or perhaps animals) who are weaker than we are and whose life chances (e.g. to live in a not unbearably oncogenetic environment) depend on us at least partly. Newborn children are naturally the paramount example of vulnerability of others as a ground for reinforcing our obligation to them. But in the case of future people, even pretty far from us, vulnerability has three further features: first, in the world of global threats we do not simply generate children who are vulnerable to the environment, as offspring of living beings have always been, but we put them in an environment made utterly more dangerous by our deeds, by our[42] actions or omissions. Secondly, if we keep acting like this (maintaining nuclear arsenals and increasing instead of curbing greenhouse emissions), we accept to inflict on them even worse *vulnera* than simply waging more conventional wars or polluting the river next door, as previous generations have done. Thirdly, we know about that, we have plenty of information and credible forecast about the consequences of nuclear war or enhanced global warming (reflective vulnerability).

Let us clarify the argumentative path we are following. As observers we watch the likely reproduction of a trust relationship between parents and children or grandchildren generations, take note of the normative potential (transmission of trust, respect for the vulnerable) that is built in in the chain, but on the basis of recently emerging threats argue that our concern, if it is to remain effective, must be extended to collective survival goods for posterity. Global challenges have also in this regard the narrow meaning designed at the outset of this book: since 1945 and *a fortiori* since the discovery of global warming in the seventies, we know that each parent generation is collectively cause of potential (nuclear weapons) and actual (climate change) damage inflicted upon their immediate descendants, and that it can now live up to the obligation it takes up by generating children only if it learns to widen its action range from the individual care for one's own children and grandchildren to the collective and political level; that is, only making the planet safe for all other generations' children will bestow full sense on the individual care itself. In plain words, since global challenges have arisen it does make pretty little sense to set apart just (as argued in *A Theory of Justice*) or supererogatory savings for future generations if we (the future as well as the present "we") do not at the same time do our best to prevent the global challenges we have ourselves created from becoming actual catastrophes.

Two difficulties arise at this point. Before we go over to intergenerational obligations, the problem of infragenerational solidarity among parents and their social groupings must be addressed, as peoples or political groups among them have so far not refrained from killing, repressing, or enslaving the children of other human beings. This is however a political rather than a moral problem, and results from the division of humankind into particular groups and polities; it has been recently alleviated by the end of colonialism and the rise of humanitarian aid and legislation. Except if one argues from a racist position, there is no theoretical problem in conceiving of infragenerational solidarity, be it as an universalistic attitude, be it as a result of reciprocal deterrence ("if you hurt my children, so I will do to yours"). Second comes the possibility that, mainly with respect to global warming, each generation acts or rather omits to act out of the so-called "generational nepotism": we understand that we are endangering the planet by not curbing greenhouse emissions, but we also know that really dramatic consequences will affect only generations far ahead of us, so that we may egoistically think that we do not need to worry for us and our children and children's children, especially that we are not bound to take upon us the costs needed for restructuring our economy and curbing emissions. This is in fact what seems to happen among the generations living at the time of writing: they have successfully addressed the ozone layer protection issue that threatens our contemporaries with skin cancer, but done very little or nothing to curb greenhouse emissions, whose consequences are less direct and recognizable. Not unlike the danger created by nuclear weapons, in the psyche of our contemporaries powerful denial mechanisms ("the warming is not so high, or does more good than evil, or its effects are so far in space and time from us or so unpredictable that we need not worry") are at work and are also fostered by the propaganda machine of antienvironmental vested interests.

In other words, global challenges have affected the intergenerational chain too: in order to give future generations the same freedom to generate and educate children in decent and improvable conditions as we and our parents did (in developed countries at least), it is no longer enough to take care of one's own children and children's children and let our contemporaries do the same. This is why at this point the mere projection of caring parenthood into the future turns out to be a fundamental, but still insufficient basis for giving firm roots to the concern for future generations. If this is to be achieved and the nepotistic escape from this concern to be countered, a deontic element must be introduced and from observers we have (or rather we choose) to turn into participants, active subjects of morality, even if what is needed is just basic morality and not an elaborate moral philosophy. By the way, this, if not other elements, should preserve my argument from being indicted for infringing Hume's prohibition to translate a fact such as the chain into a norm. In other words, if the chain is not to be broken, the danger that this may happen has to be made clear and widely known at the cognitive level,[43] while its normative potential has to be complemented and made effective by an argument telling us not to irrevocably damage the physical and socio-economic conditions in which future parents and future children can enjoy at least as much freedom as ours.

The argument must be as plain and thin, that is, free of particular philosophical positions and language, and as transcultural as possible, in order to be accepted in all political communities of polluters and present or prospective owners of nuclear weapons. It must express the elementary sense of fairness between human beings that, in the Judeo-Christian culture, has found an early expression in the *Golden Rule*, of which I shall cite the formulation given in Luke 6:31 "and as you wish that men would do to you, do so to them," although more immediately applicable for our purpose is the negative formulation given in the Old Testament "and what you hate, do not do to any one" (Tobit 4:15). The transcultural nature of this rule is visible if one looks at its likely origin in India and its presence in Confucianism.[44] Applied to future generations, it loses all elements of *do ut des* reciprocity (we are all dead when the posterity receives the benefits of our fair behavior) that may stick to the Golden Rule as an obligation among contemporaries.[45] The only self-interest we may uphold in this case is an ideal one, in the sense that we aim at preserving our moral identity as parents by not damaging future parents (more on this in chapter six).

In the given context of problems, a basic sense of fairness can work on either of two levels, or indeed on the both of them, if it is to secure the obligation to preserve survival goods for all future generations. Either we assume that on this ground each parent generation sees as its duty to avoid damaging acts or omissions capable to create on the planet conditions that would make it difficult or impossible to future parents to care in an effective way for their children, because by then the earth will be irreversibly spoiled. Or our sense of fairness tells us to redress the unbalance that their enhanced vulnerability to the consequences of our actions has created between us and future generations. If this argument is convincing, we have reached our goal: to argue and make visible that obligation in the "thinnest" possible way, introducing just the most elementary and widely shared (also among cultures) deontic moment and avoiding all more substantive and therefore precarious theories of obligation.

The trust we give ground to others to put in us[46] as well as their enhanced vulnerability to our actions/omissions, inasmuch as it reflects upon us, give thus foundation to obligations to them and generate *responsibility* towards them. Responsibility[47] implies that

1. We are able to cause certain effects and know about this causal link[48] (as it is doubtless the case with nuclear weapons, but in a more restricted way also with greenhouse gases emissions);
2. We deem these effects to be damaging and unfair to other people, because they deprive them of basic life chances similar to ours; lastly
3. We are free to change our course of action and to try to avoid or mitigate those effects.

We can thus say that taking responsibility to future generations with regard to global challenges is implied in bearing children or helping others bear them, provided everybody looks at his or her relationship to future generations of parents with a minimum of fairness. In other words, responsibility presupposes the existence of moral actors, that is, actors endowed with the minimal structures of morality such as the Golden Rule. Which forms responsibility should take[49] and in particular whether or not the "precautionary principle" can be regarded as its leading development are not the problems I am going to take up in the following, because this is unnecessary to our main argument, which is aimed at finding foundations to an obligation towards future generations, rather than stating how it should be implemented.

This is all relatively plain on the side of morality, but far more complicated when it comes to action, policy, and actors. That responsibility for future generations is implied in the basic anthropologic and moral structures of *animal humanum* ushers in the task of making explicit what we deem to be implicit, to draw political consequences (i.e., actions and motivations to act) for humankind from what we are used to doing in our private and family life.[50] Taking a moral obligation into the field of political action is a difficult task because the two spheres have different codes and criteria. On the road from one sphere to the other the actor itself "we" change his traits: while all human beings are equally addressees of a moral obligation, there is at first no such "we" in politics, where rulers and the ruled, members of the elite and common men, full citizens and resident or illegal aliens are separated from each other. This is just an announcement of problems that will keep us busy in chapter seven, but the road is still long before we can take them up. The world in which we are confronted with global challenges is far too complicated to be adequately understood along the patterns of normative alternatives between just and unjust actions. Alternatives like this are embedded in broader frameworks of situations and processes, which will be the subject of our next chapters, and also imply a shift in the moral vocabulary towards a redefined notion of humankind that we are now going to elucidate. This I do not just because I still have to draw conclusions from our reasoning about human beings and future generations, but also in order to clear the way in the moral terrain for what we shall discuss in chapter seven concerning the political profile of that redefined notion.

However, before we go over to elucidate "humankind," it is useful to reply to an objection that can be raised[51] against the priority of survival against justice. Survival not restrained by justice can mean nuclear peace and reduced emissions under a worldwide tyranny of Hitlerian or Stalinist make. Now, while I still stress the importance of not justifying the choice of survival goods on the basis of a particular conception of value, history, and society, the normative (chapter five) and philosophical (chapter six) motivation I am giving for this choice is such that survival, chosen on these grounds, contains enough vaccine against such an unqualified and "zoological" survival that can be chosen regardless of the moral and political regime in which it is embedded. There is also a counter-objection coming from political science: by its own dynamics a worldwide tyranny that realizes Kant's nightmare of a "despotism without a soul" is bound to generate more strife and more pollution rather than to stabilize peace and to keep the air clean. Arguing in favor of the choice for survival regardless of the several conceptions of justice does not characterize the hopefully resulting survival of humankind as barren of any justice.

4. HUMANKIND UNDER MAN-MADE GLOBAL THREATS

Since the beginning of the Western reflection[52] on *genus humanum* in Roman philosophy the several versions of it can be grouped around two main (and simplified) notions: one is quantitative or rather regarding the humans as species, the other qualitative, regarding the very nature of human beings, which they should make flourish in their lives individually as well as across generations.[53] The first notion always had an anthropological rather than a moral relevance, and was used to bind together civilized peoples and "savages," Christians and non-Christians ever since the Europeans were confronted with the human groups and cultures discovered by their own explorers. The second and prevailing one, already seminal for the Stoics, had its peak between Humanism and Enlightenment,[54] becoming not only the goal to achieve, but also being merged with the *Artbegriff* and representing the transgenerational dimension in which first the intellectual and moral potential of the human being can develop and come to self-realization: this twin aspect is mirrored in the ambivalence of the term "humanity," meaning at one time humankind and the moral quality of being or becoming truly human. This version of *Menschengattung* is present in Kant's philosophy of history as well as in Marx' philosophical anthropology that can be read in the *Grundrisse* manuscript. The criticism of rationalism and bourgeois culture developing in the second half of the nineteenth century (Nietzsche, later Spengler, to whom humanity was just a "zoological notion") put discredit on any idealistic and progress-bound notion of humankind, and even more so did the two world wars. The postmodernist disdain for all holistic dimensions has then brought the unhappy destiny of "humanity" to the last consequences.

Much to the surprise of postmodernists, it seems now that the two global challenges that have broken up modern order have not just reinstated the concept of humankind, but even given it a non-notional existence for the first time in history. The political aspects of this turn shall be discussed later in chapter seven, while we are dealing here with the philosophical and moral aspects that may be relevant for politics. At the end of § 2 in this chapter we have already stated the necessity to complement the foundation of normative moral theories by adopting a meta-imperative that tells us to do what we can in order to allow for the continuation of human life, which includes civilization, as the latter is the indispensable setting and presupposition of human life even from the "zoological" vantage point of human anthropology. In § 3 we have seen that future generations or rather their existence as far as it depends on our deeds[55] is an essential complement of our life, if the minimal order on which it is based is to be preserved, instead of being torn down by self-defeating actions or omissions such as those that keep our kind and civilization in existential danger. We shall soon see in chapter six that this holds not just from the normative point of view, but also for reasons of meaning.

In the moral dimension the new significance of the concept of humankind does not come from the reinstatement of its qualitative version, which would be awkward to strive for in times of post-metaphysical and post-progressive thinking. That version only makes sense in an overarching narrative, a *grand récit* that sees in progress and self-fulfillment the marks of humankind's development. What happens instead is that the quantitative dimension of humankind, the *Artbegriff* itself, complemented by future generations, achieves moral significance, and this because of the unprecedented situation in which the members of the kind have become themselves responsible for its annihilation or preservation. This raises two questions: whose responsibility? and also: why not nihilism instead of responsibility?

It is evident that responsible for the huge increase of destructiveness in warfare and the blind use of resources that have led to global warming were not the former or the present generations in their entirety, but the decision-makers and those who took advantage of those developments: ruling classes, bureaucracies, powerful and imperial countries. This responsibility is mitigated but not cancelled by the secondary share ruled classes or countries may have had in those benefits or by their inability to advance alternative but innovative policies[56] that might have modified the direction originally imposed by the rulers. More important is another consideration, based on the physical features of global challenges: there is no way of separating in the future the responsibility of the perpetrators from the responsibility of the innocent or the less guilty. Emissions from a Chinese peasant family small car will increase greenhouse gases in the atmosphere not less than the corresponding fraction of fuel burnt by one of the several cars owned by a middle-class European or American family. Vice versa, the enhanced global warming will hit both rich and poor, affluent and underdeveloped, even if their chances of sheltering themselves are not equal. Not unlike with nation-states, the fact that we are all in the same boat, whether or not we freely decided to board it, makes us all members of humankind responsible for its future, even though with different burdens of guilt for past actions and different amounts of responsibility for what is now to be done. In principle, there is now a non-rhetorical moral of humankind, bound together by attributable guilt, and this could be reinforced by infra- and intergenerational *solidarity*, should men and women choose to address global challenges not only out of a sense of common threat, but also because out of responsibility they develop feelings of reciprocal belonging. Ironically, this would happen exactly at a time when (and contrary to the fact that) "humanity" finds little or divided acknowledgment among philosophies and cultures. Whether or not there can be an equivalent political community will be a main topic of chapter seven.

As already noted above in § 1, as an alternative to holding ourselves responsible, *nihilism* is significant only morally, not politically. But this qualification does not make it intellectually insignificant. First, we need some conceptual clarification. Nihilism in contemporary philosophy means a radical denial of, or a radical skepsis about the basic moral and civilizational beliefs shared by our society, in the way of Nietzsche's "upside-down change of all values" (*Umwertung aller Werte*), or of the rebellious defiance shown by Bazarov, the nihilist young doctor in Turgenëv's *Fathers and Sons*. Other, less encompassing examples can be found in Hume's characterization of the not-unreasonable passion that can let me "prefer the destruction of the world to the scratching of my finger" (Hume 1739, 463) or in G.E. Moore's statement that, as the pessimistic view that "the existence of human life is on the whole an evil" cannot be proved or refuted conclusively, universal murder leading to the extermination of the human race should be ruled out only because it is impracticable as a means (Moore 1903, 205–6); besides, let us not forget the passage by de Sade quoted in note 5. Nihilism is a conscious or even well-argued intention to subvert all values up to making us indifferent towards or even eager for the destruction of humankind. But this is not what comes up with the global challenges, which is the potential to destroy the world of humankind as the involuntary or even counter-intentional outcome of a real development initiated, but not controlled by human beings. On the one hand is theoretical value of nihilism, on the other hand annihilation, or in other words practical, perhaps unintentional nihilism.

My thesis is first that philosophy has much too much taken up nihilism and overlooked annihilation, which it has hardly thematized. Secondly, that any attempt to derive the latter from the first is flawed and denies annihilation as a problem of its own. To prove the first assertion a look into the literature on nihilism is sufficient: this is examined in all possible shades of philosophical denial of truth, reason, value, and obligation (against cognitivist theories of morality), never as effective possibility of human life ending in *nihil*. On the other hand, especially in Christian images of modernity, a conventional wisdom can be found that ties the doomed perspective of humanity in recent times, including nuclear Armageddon, back to the collapse of religious faith and traditional morals, to secularization and relativism. This overlooks the fact that the cultural environment in which nuclear science and technology came of age and the belief in industrial civilization flourished was hardly shaped by relativistic or nihilistic attitudes, which played no role in the life and mind of people such as Albert Einstein and Henry Ford, the founding fathers of twentieth-century physics and rationalized industrialism, or Franklin D. Roosevelt and Harry Truman, the U.S. presidents who built and employed the atom bomb. There is indeed no credible scientific attempts to prove these ties according to the standards of historical inquiry. It was not value nihilism that set the devils of annihilation free. Overstating the power of philosophers and doctrinaires in shaping real developments in civilization or mistaking a real threat for the mere consequence of philosophical beliefs[57] may please the self-esteem of intellectual elites, but also creates a bizarre disproportion between suggested instigators, the nihilist *literati*, and effective, nuclear or environmental, threats. In the next chapter we shall enter other pathways to tentatively find out about the roots of our eventual annihilation.

Yet, after ruling out a historical link between nihilism and annihilation, we are nonetheless left with the philosophical question if a nihilistic indifference towards the normative attitude of taking annihilation seriously and acting correspondingly can be countered. Oswald Hanfling put it in the way that the existence of humankind cannot be preferred to annihilation on grounds other than the respect for the wishes of those who, now or in the future, care about its conservation (a reason of elementary fairness, as I have said before); or because its conservation with its unique features matters particularly to us as humankind members, which sounds like a pretty circular justification, or rather no justification at all, as it amounts to the pure (according to Kant) act of will *sic volo (vivere), sic iubeo*. Other reasons such as God's will or the respect for biodiversity or giving the human potential the chance to develop in its own dimension (the kind, not the individual) are too lopsided or controversial (for some environmentalists, biodiversity rather requires our extinction) or require to assume and to prove too much in advance. So also does the Kant-Marx argument on the kind (*Gattung*) in whose dimension only the individual's potential finds fulfillment, so that we cannot but want its preservation: one can as well think that what he or she has achieved and experienced in his or her life is all that matters, and it is in any case awkward to refer one's own colorful life to such an abstract entity as the kind or race. At most, we can turn philosophers and interpret what we effectively do (to generate children and to attempt at giving some of our legacy over to them, beyond protecting them and caring for their future) as projection of our individual world into the *Gattung*; but this an "as if" interpretation, and scarcely the only possible one (we can for example find God more appealing than the kind).

In conclusion, to the question raised above the answer is positive in moral terms, if we look back at the attempt made in § 3 to found the defense of humankind from global threats upon the normative implications of the elementary structures of human life; it is now evident that I have kept the argument as thin as possible in terms of assumed values and principles also with the aim to present the smallest possible target to a nihilistic shifting of values. This is yet not to say that defending the life of humankind is a value-free move or that the value upheld in this way is contained in the mere facticity of generating children. In order to endorse that value we must *decide* not to act like Saturnus, the ancient god who devoured his children, or like other less bloody but careless parents; we must stand up to the moral challenges we ourselves choose to be confronted with when becoming parents. What we have found out in this last section is that, if we do so, we accept an anti-nihilistic responsibility not just for our children but for the continuation of civilized human life.

In psychological terms, the answer cannot be as positive: no theory of morality will be ever capable of eradicating the possible emergence of personalities whose overblown Ego can come to approve of the coming catastrophe or regard the moral arguing about it as pointless, because in an aesthetic approach they find images of cosmic doom beautiful. It is true that destruction appeals somehow to our aesthetic sense: otherwise reasonable citizens and commentators have found aspects of beauty in the picture of the

Twin Towers being torn down in fire on September the 11th, 2001. This is not the point, but rather the *èpater le bourgeois* attitude that grants preference to this "beauty of destruction" against the moral significance of humankind's life being fully disrupted. It should be perhaps added that the other pattern of possible approval of or indifference to the destruction of humankind or civilization, which Hans Blumenberg (1997) has famously described as "ship wreckage with spectators," does not apply in the case of global threats, when all are in the same boat and there is no separation between victims and onlookers.

With nihilism and annihilation we have got to the point where moral reasoning about future generations and humankind goes over into reflections regarding the meaning that we attribute to our life and death both as individuals and as a kind. This is the first issue we are going to meet with in the following chapter, and from it a long road will lead us to an overarching philosophical assessment of significance and origins of the global challenges.

NOTES

1. §§ 44–45.

2. An influential volume was also the reader *Obligations to Future Generations*, see Sikora and Barry 1978.

3. This is a first emergence of the identity problem, see below.

4. For the present debate see Sagan and Waltz 2003; the best known document of realist fantasies of "more nukes, more security," remains Mearsheimer 1990.

5. To my knowledge it is only in de Sade's work, and even there in just one passage (a speech by Dolmancé, the master libertine, in *La philosophie dans le boudoir*, seventh dialog, Sade 2004) that the idea of possibly burning the entire world if nature commands us to do so has found a philosophical formulation; cf. also Henrich 1990, 127.

6. To be clear: this is a short for the obligation (however argued) to do our best as to eliminate the present and future risk of nuclear war, which does not necessarily entail the so-called "elimination of nuclear weapons," a particular policy whose pros and cons shall be discussed in chapter seven, and which competes with other policies addressing that goal.

7. On the attitude towards nuclear weapons in different cultures and religions see Hashmi and Lee 2004.

8. Cf. below § 4.

9. By this kind of unfairness, noticed also by Kant, Herzen, and Berlin, "those who live later profit from the labor of their predecessors without paying the same price," Rawls 1971, 291.

10. The forecasts regarding nuclear winter or autumns made in the eighties have not been put substantially in doubt up to date, they are simply less talked about since major nuclear war has become politically only a very remote possibility.

11. These topics are to be the theme of chapter seven. The difference of this perspective from the anti-proliferation regime, in which at least the "official" nuclear powers (but perhaps also the three, Israel, India, and Pakistan, which have successfully proliferated) are entitled to preserve their arsenals, is fairly evident. Let us however remember that the original Non-Proliferation Treaty of 1968 tried to entice non-nuclear powers into signing with the promise that the nuclear powers would make steps toward disarmament.

12. Even if they do not know to which generation they belong, see Rawls 1971, 292 (§ 44).

13. Brian Barry and Robert Goodin, according to Dobson's account of this debate, see Dobson 1998, 119–21.

14. This suggestion reflects a pluralistic approach to moral theory as it was unfolded by Charles Larmore (1987), Steven Lukes (1991) and for nuclear ethics by Dieter Henrich (1990), cf. Cerutti 1993, 403. To be sure, pluralistic does not mean eclectic.

15. Shalit 1995, 54–65.

16. Only Apel has made an attempt to reconcile the ethics of discourse with Jonas's ethics of responsibility, yet his assumption of a "potentially unlimited community of argumentation" which overruns time limits and creates a reciprocal responsibility between generations (Apel 1988, 179–216, especially 201) has little grasp on the challenges we are here discussing.

17. Quoted in Dobson 1998, 10. The idea of an intergenerational contract is best explained by its harshest (Shalit 1995) or moderate (Laslett and Fishkin 1992) critics.

18. This was indeed the key in which I used the notion of a contract between generations in Cerutti 1993, 93, but I have come to believe that a metaphorical use is not justified in this case; more fundamentally, the use of metaphors, though unavoidable, should be kept at the lowest possible minimum in political theory.

19. Rawls 1971, 11.

20. Rawls 1971, 289.

21. Rawls 1971, 128. More exactly, it is the persons in the original position who are not thought of as having "obligations or duties to third parties, for example to their immediate descendants."

22. Rawls also specifies that capital is not only factories and machines, but also knowledge and culture, techniques and skills (Rawls 1971, 288).

23. This expression from Locke's *Second Treatise* (Locke 1690, 132) is a *topos* in the literature on ecological ethics.

24. Reference is to *Political Liberalism* and *The Law of Peoples*, in which, however, little or no attention is given to the problems of future generations and only marginally to nuclear war.

25. These two patterns are indeed both present and intertwined in the case of nuclear weapons.

26. I have summarized this debate in Cerutti 1993.

27. Parfit 1984, Part Four.

28. Parfit 1984, 388.

29. Rawls 1971, 62. Rawls also mentions natural primary goods such as health and vigor, intelligence and imagination.

30. Kavka 1978, 191.

31. Or primary goods, if we want to stick to Rawls' language usage.

32. See above p. 54.

33. This is the tradition of modernity, to which this approach has been so far seminal. In a still metaphysical key (transition from time to eternity and its significance for morality) Kant has speculated on the "end of everything," cf. Kant 1794.

34. "Bomb them [the enemy] back to the stone age" or "to bomb humanity back into the stone age" were parts of the Cold War idiomatics.

35. The few exceptions, all in the fifties, have been Anders' and Jaspers' writings on the atom bomb and in a wider sense Heidegger's essay on the essence of technology.

36. On trust as a general sociological category see Luhmann 1979 and Gambetta 1988.

37. I shall not explore in this book how far this notion resonates with the ethics of care developed by feminist thought.

38. Legal and moral writers such as Pufendorf, Mills, Sidgwick, and Hart have similarly justified the obligations of parents and the rights of children, cf. Goodin 1985, 81–82. I shall not take up the notion of rights (for a discussion see Baier 1984, Shalit 1995) and use it only in a generic sense.

39. The idea of a chain (of love, not of trust and obligation) across generations has been advanced by Passmore 1974.

40. I am saying beings and offspring instead of men, women, and children, because we do not know by now how human beings may be reshaped by bioengineering and identified in five centuries.

41. I owe my attention to this notion to Goodin 1985. It implies that we acknowledge a causal link between our actions and those affected, as I shall highlight below with regards to responsibility.

42. "Our" includes in this case the present and the previous generations, in other words of humankind (past and present) as cause of damage and (as far as we have become aware of this causal link) perpetrator of evil against the humankind due to live in the future.

43. The British government plans to introduce the topics of climate change and sustainable development in the education of schoolchildren, see Garner 2007.

44. The wide transreligious diffusion of principles similar to the Golden Rule has been demonstrated in the unknown, but useful Ph.D. thesis by Leonidas Philippidis (1929).

45. On the moral problems implied by the Golden Rule as a logic of equivalence see Ricoeur 1990. I am indebted to Stefano Petrucciani for critical remarks on the role played by the Golden Rule in my argument.

46. Directly, our children, or indirectly, future people in general. In this sense we are in principle all bound together by the "general fairness that cements mankind," as Locke put it (see Laslett and Fishkin 1992, 32).

47. Hans Jonas (Jonas 1984) had the indisputable merit of being the first to work out the time-dimension of ethics and to argue the primacy of responsibility over hope,

the leading principle of utopian Marxism. But his conception of responsibility has taken little note of fundamental shifts in philosophy such as the linguistic turn and the paradigm of intersubjectivity (Habermas); his metaphysical argument for the continuous existence of humankind as a categorical imperative and the "ontic paradigm" (Jonas 1984, 235) which simply translates the "is" of the newborn into the "ought" of our caring for him or her do not stand critical examination of their philosophical roots. This circumstance along with Jonas' lacking sense for the difficulties of talking humankind into a new sense of responsibility as well as for skeptical counterarguments make *Das Prinzip Verantwortung* a manifesto of remarkable motivational power rather than a state-of-the-art justification of the obligations to future generations. This is why I shall scarcely make reference to it.

48. The relevance of (reflexive) causality to the establishing of obligations towards members of the human race unknown to us as posterity is highlighted by Dobson 2006 in a context of "thick" moral cosmopolitanism.

49. Relevant can be in this sense the distinction of self-responsibility and other-responsibility made by Care 1987.

50. Similarities and distinctions between the responsibilities of parents and statesmen are discussed by Jonas 1984, chapter 4, § 3.

51. I am grateful to Alessandro Ferrara for drawing my attention on this point.

52. There is no consensus as to whether this reflection can be predated back to the Greeks, see Baldry 1965; *contra* Bödeker 1980.

53. Bödeker 1982, whom I owe this clarification, speaks of *Artbegriff* and *Zielbegriff*.

54. See Schlereth 1977; for Locke mankind was *one* community before being disrupted by the corruption of degenerate men.

55. And not on cosmic events such as the cooling down of the solar system.

56. Instead, unions and working class parties in the West have longtime shared with the industrialists of their countries the limitless priority of an environmentally short-sighted notion of growth.

57. For this reason I disagree from Ungar 1992 who, however, rightly raises the issue of the psychological components of the mentality favorable to nuclear dominance.

Chapter Six

Meaning and History under Global Threats[1]

We have so far examined the chances to justify an obligation to act with the aim to free the world from the two impending global threats. But while obligations are founded in ethical arguments, the ethical foundation is not exhaustive of all that can be said about them. They must as well make sense to the actors; ethics must be complemented by meta-ethical discourses. The meaning of global challenges and our reaction to them for our sense of life and death, both individually and collectively, is therefore the subject of § 1. The meaning we put in our life experience and in the emotions such as fear that come across it can indeed be interpreted not only phenomenologically, but also historically. This is why § 2 will ask how far global challenges mean the end of modernity, or a major cleavage in it, thus extending into a wider, while not only political, frame of reference, a question we have already put to debate in chapter three. It is in this framework that another question, which has surfaced several times in this book, shall be discussed systematically in § 3: how much and why modern science and technology have contributed to generating the two global challenges we are dealing with.

1. THE MEANING OF HUMAN
LIFE UNDER GLOBAL THREATS

Meaning is a common word, which is often used without much thinking about what it means. It is therefore helpful if we first state the meanings of "meaning" that are excluded from our context: I do not relate to any "objective" meaning, the meaning of our life in a world order, because I do not think an ontological order can be assumed, nor can an easy consensus be found on it.

Nor do we refer to the meaning of life according to a particular conception of the good, because particular conceptions are too narrow a stage for the all-human range of problems we are here confronted with. Neither do we regard as relevant to our argument the question whether meaning in general, in which something can be meaningful also from an egoistic point of view, must be separated from significance, which can be obtained only if we go beyond our personal well being (cf. Singer 1992, 114–18). As in the last chapter, we make our start with the thinnest possible basic assumptions.

What is then the notion of meaning we refer to in these pages? We think of it like a grid in which we put events occurring in our life and acts that we perform; in doing so, we put them in a cognitive as well as value-related relationship to an order (such as a life plan[2]) or a narrative, which can be made of memories and/or future perspective, as far as projects of liberation or salvation come into play. Without developing a storyline in which we can recognize ourselves, there is little chance of building both the individual and the group identity. In the cognitive aspect of the meaning relationship events and acts become understandable, dropping their pure factness and finding a causal or functional or intention-related explanation; in the value-related aspect[3] they become acceptable, helpful or damaging that they may be to us, because they are then illuminated by our positive or negative value-judgment. Examples of values or rather valuable conditions we may be eager to attain are life itself, whether or not we think of it as something which is in any case worthwhile to be lived,[4] dignity and self-esteem, public recognition or glory, duration in power or wealth, in other words what Thomas Hobbes knew as passions, granting them an important role in his political anthropology.[5] These values and the passions tied to them are vital remedies to the meaninglessness of our individual death, as the grid of meaning which we develop around them helps us overcome the otherwise overwhelming sense of absurdity arising from the contrast between the life we are living with hope and desire on the one hand and the brutish matter-of-factness of a death as inescapable as possible at any time.[6]

As the meanings we try to give to our individual lives strive to overcome the absurdity and unacceptability of death, they need to do so in a meaning-providing (*sinngebend*) structure that overcomes a life span and reconnects individual life to past and future.[7] The relationship between life and death has not been always shaped in the same way in Western culture, as the discovery of the significance of death first required a completed cultural process of individualization and the invention of non-cyclical, linear time (see Choron 1963). The construction of meaning as a structure in which we come to terms with a world in which we know we have to die has been a specific achievement of modernity. In this era earthy life has acquired a value of its own, and does no longer or not necessarily depend on transcendence in order to be accepted and made sense of, nor does it resign in front of natural catastrophes[8] or the perspective of the death of the solar system, and more generally of a world described by natural science as having as such no meaning, because—unlike in the creationist view—it does not have any purpose, and matters only as the object of our *curiositas* and our will to acquire unlimited power on it. In the modern construction of meaning, which cannot be more thoroughly discussed at this point, a major role has been played by the process that consolidated into institutions our knowledge and mastery over both the natural world and the human subjects of sovereignty[9] and also enshrined our deeds in a philosophy of history and progress overcoming the individual's death. This is only a case of the general requirement for meaning to be made credible and transmissible by being shaped as a narrative (not necessarily a myth!) whose syntax is provided by symbols.

Now, these modern tracks for the production of meaning have begun to collapse under the appalling strains put on it by the two world wars, the totalitarian regimes, Auschwitz, and the *gulag*. This happened on the background of communitarian bonds being torn down by capitalist modernization, individual biographies being subject to the anonymous necessities of the "iron cage" of modern work (Max Weber 1905, 181) and of the highly bureaucratized mass society built around it; all processes that amount to a twin loss of freedom and meaning (*Sinnverlust*).[10] On this social and cultural background and as a political conclusion of the first half of the twentieth century, the A-bomb has given the final stroke to the destruction of the common sense asserting the reality of human progress by means of science's advance, and after this any progress-based philosophy of history has lost sense;[11] the later environmental shocks have been just an unexpected reminder. The meaning-eroding mechanism does not lie simply in the circumstance that, if the permanence of human life after the individual's death is threatened, there may be then no men and women at all who can substantiate the meaning-creating projections[12] we make on future generations. This is trivial, but it is perhaps not useless to speak it out. The point is that the possible disappearance of human civilization would come about by this kind's own hands, while the fact that the perpetrators would be actually some bad guys rather than the whole kind would not make any real change. It is a matter of moral symbolism, a very substantial force in morality: if humankind kills itself or poisons everything its future members need to survive, to exercise fairness or *a fortiori* solidarity or sympathy, let alone universal love, would simply make no sense and could not be seriously made the case for. Humanity as the totality of human beings would disavow itself as a moral subject[13] and even as a credible rational actor, since the appealing validity of instrumental and/or strategic

reason would not survive such a destructive outcome to the Baconian intention of asserting man's predominance over the world. Along with Bacon,
among the losers would be Hegel, one of the standard-bearers of modern Reason in its encompassing project to redress not just the natural world, but society and history as well according to itself, or—as he put it with reference to
the French revolution—to the *nous*.[14]

However, the modern settings in which human generations have been able
to give a meaning to their lives can be uprooted by global challenges also on
a more basic level, which comes closer to everybody's common experience.
Already as a sheer possibility, nuclear war affects the life of the presently living individuals; along with man-made climate change, which is not a sheer
possibility but a growing real burden, it could make the life of future generations "nasty and brutish," to put it in the terms of Hobbes' description of the
state of nature. Unlike fifty years ago, the vast majority of the generations
now living on the planet knows about this prospect, which involves not only
the people of the far future, but our children and grandchildren as well. At risk
are thus the chances to combine our experiences and life plans into the meaningful unity of one's own autobiography,[15] because the stability of both its
physical and intergenerational pre-conditions is no longer given. Poorer is
also the chance to keep the always-present threat of chaos at bay by entering
and supporting a mental and institutional order, be it social, legal, or political
(cf. Berger and Luckmann 1967, 103). This was indeed the task which early
modernity assigned to mathematized science,[16] natural law, and the political
construct, the Leviathan, based on it. How can we indeed still rely on this
chance, if in both science and politics human agency is discredited because of
the perverse effects deriving from those pillars of modern order? By the way,
these effects put under stress another feature of modernity, secularization,
which has to compete with religion in satisfying the human "aspiration to exist in a meaningful and ultimately hopeful cosmos"(Berger 1977, 79). Hence
the return in the West, but not only here, to sundry religious or magical beliefs, as a renewed form of self-reassurance, often with a wide presence of
end-of-the-world images and expectations.[17]

It is not only that the individuals have less chance to shield themselves
from chaos, because human agency is discredited by the self-defeating outcome of its modern endeavors. A further difficulty lies in the question of how
to give the individual biography the meaning that it used to receive from its
connection to the life of predecessors and posterity, if this is not certain to
come into being and the former's intentions and hopes have been so radically
disregarded?

Against this picture of the relationships between present, past, and future
in the meaningful build-up of personal and social identity an objection of ob-

solescence could be raised: in this reading, holistic dimensions such as life plan, family life, cumulative wealth, and well-being for oneself and the near others are already outmoded, self-realization is rather sought by shifting life plans, flexible and highly adaptive patterns of behavior, while patchwork has replaced unilinear build-up in projecting and narrating one's own biography. This is partly true, but grossly exaggerated if it is assumed to hold for the majority of the youth in the developed (not to speak of the developing) countries and for most of a person's lifetime, since the need to stabilize one's own life perspectives does not fail to come up at some point even in the most postmodern life history. On the other hand, the presumptive new settings for biographies should be seen not only as a manifestation of a new liberty and autonomy, relying on less substantive or thick principles and structures, but also as the result of destabilizing factors: less certainty and security in the social link of work and personality,[18] less reliance on the protection given by the state against external and environmental threats. Fear of nuclear disasters, whether deliberate (another Hiroshima, or a terrorist "dirty bomb") or unintentional (Chernobyl), or environmental decay play a role in defining the frame within which individuals become aware of the world they are born in and must take note of whichever orientation they may want to give to their lives. This has been shown in the *histoire des mentalités* of the generations that have lived in nuclear hope and fear between 1902 and 1986,[19] in the psychoanalytic literature on "the psychological fallout of the nuclear threat,"[20] and in social-psychological studies on the impact of the image of extinction on family relationships.[21] Nuclear fear has subsided with the end of the Cold War, as the world's public opinion gratefully acknowledged that there was new real ground for solace and less need of the denial mechanisms that in the worst decades of the nuclear arms race had been a deforming as well as protective defense against too sharp and unbearable a consciousness of the impeding threats. That fear has been possibly replaced by the fear of terrorism, spontaneous (in the aftermath of September the 11th in New York, March the 11th in Madrid, July the 7th in London, and everyday in Baghdad) or artificially inflated[22] that it may be; or by other forms of fear, such as those tied to what is called "urban insecurity."

In general, we know little about this evolution: empirical research on these topics is infrequent and fragmented, they seem even out of the research agenda, because clear-cut and overwhelming causes of fear like the nuclear challenge between superpowers are gone or have retreated, or because the theoretical understanding of the links between fear, meaning creation or destruction, actual event and policies is underdeveloped, or unappealing to established academic research. The present lack of empirical research must be taken into account, but is not necessarily a sign of nuclear and environmental fears being now irrelevant to the meaning we give to our life as individuals and humankind as well. A philosophically informed diagnosis of our time such as the present work cannot replace that lacking research and should avoid overstating its own cognitive strength, but can keep the problem open and help stimulate a subtler interdisciplinary research agenda. For example, narcissism as the inability to connect emotionally to the object-world and the other human beings could be examined at one time as the attitude that hinders people from acknowledging and working out their own fears and the self-defensive response to the otherwise unbearable fear of total destruction. How has this attitude, first defined in the seventies by Lasch and Sennett, evolved in these decades, particularly under globalization and worldwide terrorism? Is it still a workable pattern in order to understand how shared meanings are worked up or dissolve? In any case it does not seem to have created to a remarkable extent a new independent social-psychological type, but rather to be a possible component (a temptation?) in the complex and mixed Self of present days.

Whatever the answer, we can in any case conclude that even in the case of postmodern women and men, to find a meaning to one's own life has become more difficult in the presence of global lethal threats, and that addressing those threats with awareness and action can help create a new common sense and somehow restore a meaningful relationship to our posterity. If this is true, it can provide a more favorable cultural environment to the moral obligations to future generations we have discussed in chapter five with regard to their justification. Why? Normative morality is not self-sufficient, but must make sense in the broader reference system of the actors: this is how I would like to describe a problem that is otherwise known as the problem of the relationship of morals (what is just) and ethics (what is a good life). In moral theory we can successfully argue that we have a duty (or metaimperative) to stand up for the preservation of human race and civilization from self-inflicted destruction. But this reasoning can motivate us and direct our course of action only if we find that preserving our race makes sense to us and we have an (ideal) interest in it. This formulation is less engaging than the insistence on an ethics of the good necessarily complementing the criteria of justice and stands clear of any communitarian aftertaste when it comes to indicate the limits of morality. Pointing to the salvaging of the preconditions that make it possible to give meaning to our life on this planet is a wording not as demanding as an "ethics of humankind."[23]

Two remarks on this conclusion. I have directed my argument primarily towards the basic preconditions for human life to be meaningful such as its being embedded in the space between the world before and the world after us[24] rather than insisting on the religious and philosophical conceptions that claim to create meaning. This I did for two reasons. If the arguments in favor of taking the global challenges seriously want to be as universal as the challenges themselves, they must, as far as it goes, cut down their ties to particular cultures. They must in principle be able to apply to people who share with us Westerners the eventual condition of victim, but not the configurations of the Self that we receive from our particular and highly complex cultural history. As an empirical assumption, we know that the majority of those dwelling on the planet have many other and more stinging worries—from famine to disease, from war to local economic growth—to think about than nuclear weapons and climate change.[25] Hence the focus on facts, values, and meaning structures that are possibly common (with regard to global challenges) to the majority of our non-Western co-dwellers, such as children-bearing and family life.

This very same preoccupation with a trans-cultural approach is justified also with reference to the knowledge cleavage within societies. We should be cautious enough not to take the theories and narratives, from nihilism to postmodernism, from utopias of progress to those of race, whose emergence or obsolescence among intellectuals mark the shifting of meanings given to our philosophy of life, for the main or exclusive indicators of what really counts in the consciousness and motivation of common men and women. There can be intersections, and in the last two or three centuries sometime theories have become mass movement; but there is no necessary coincidence, and often a crass discrepancy exists. Intellectual cultures, the explanation and justifications they offer to the actors are often the only hint an era passes on to us in order to understand its course, and we incline all too easily to pick it up as a truthful mirror; but a different type of enquiry (social history, *histoire des mentalités*, ethnological studies, and more) should provide us with the evidence that would come from a closer look at *all* actors. Unfortunately, these enquiries are hardly made, because there is little readiness among researchers to put reaction to global challenges on the empirical agenda, perhaps also for the lack of a convincing theory that focuses public attention on them.[26] This is said only to justify the repeated warning that we know indeed fairly little about the real state of consciousness of the people in the several continents and cultures under globalization and global challenges,[27] that we only cautiously try to cart poorly known territories, and the relevance intellectual cultures give or deny to these unprecedented settings in human history is not necessarily the same as we may find among common people and large movements. What the latter have in mind is indeed to be taken into account, if we want to make good with the definition of global challenges as those that hit everybody and can only be countered by everybody's mobilization: this implies the moral and political relevance of the perception that everybody around the world has of those threats and the meaning he or she gives to them.

2. POST-MODERNITY IS NOT POSTMODERN

It is not the first time in this book that we address the issue of how far modernity has been disrupted in its essential structure and what may come after it. In the first round we have looked at the specific contribution that the most thoroughly examined global challenge, that is, nuclear weapons, has given to the demise of the modern *political* order, already undermined by the upcoming of ideological and total war. This contribution seems to have a more substantial weight in pushing that order to an end than other factors such as economic globalization affecting sovereign statehood or ethno-religious terrorism assailing basic tenets of modern civilization. But we have further seen that nuclear weapons as well as man-made climate change seem to put in question ethical and symbolic (e.g. meaning-related) presuppositions of the order in which we and our forebears have lived for the past several centuries after the end of the Middle Ages. In the second round we have thus to tackle the question of what happens to *cultural modernity* under global challenges.

In doing so we must preliminarily answer the question of the similarity or dissonance of this assessment of modernity with the current postmodern standard view, from Lyotard's inauguration of postmodernism almost thirty years ago to more recent elaborations by Bauman.[28] Dissonance prevails over similarity, as I would suggest that global challenges give leeway to a non-postmodern view of post-modernity, but let us start with similarities.

Global challenges, as I noted quoting Bobbio, give a major and indeed definitive blow to the *grand récits* or (progressive) philosophies of histories[29] or metadiscourses on which modern Western humanity relied to make sense of its epoch. An animal race whose remarkable evolution may nonetheless end in self-inflicted self-destruction cannot really explain and praise this evolution as a secure ascent from lower to morally higher phases. Yet what collapses is such philosophical and evolutionist description of history, not necessarily the values or aspirations, such as those of the Enlightenment,[30] that may have been raised in one of those phases. For them, a purely normative foundation remains obviously available, even without any correspondence in the historical evolution. This on the other hand is not the mere heap of evil that some nihilistic or postmodern caricature wants to reduce it to, but contains also clear elements of improvement of the human condition, both in biological and moral, legal, political terms—just to name a few of those elements, from the prevention of genetic transmitted disease to *habeas corpus*, general education, and equal rights for women. A contradictory development, between reduction of evil and its limitless enhancement, is one leading feature of modernity, or rather this contradiction, present in other epochs as well, has experienced in modernity a peak in its intensity that is expressed by the unheard-of amount of destruction encapsulated in the global challenges. I have already made several times reference to the seminal account of this contradictory development given by Max Horkheimer and Theodor W. Adorno in terms of a dialectics of Enlightenment sixty years ago,[31] but I am not going here to be more precise about my deep debt to them or the revision I would like to propose.

A further point of contact about this book and the postmodern standard view is epistemological: modernity and especially its end do not fit a systematic conception or a monocausal explanation of them. This is why for example I do not try to tie up my account of modernity's present troubles with the account of modern evolution as capitalist evolution. Rather, the nuclear and the environmental threats have been a major evidence disproving the agency of a single hidden overarching mechanism (the normal capitalist exploitation of the labor force) as the root of all social and political processes.[32] This holds *a fortiori* for metaphysical or post-metaphysical attempts to derive the present destructiveness from a fundamental philosophical posture, be it nihilism, as we have seen, or Heidegger's *Seinsvergessenheit* or whatever other simplistic and pretentious catch-all formula philosophers may contrive with all due contempt for the intelligibility of their accounts on the basis of common experience and the interaction with non-philosophical, for example, scientific or historical explanations of our world.

These are no tiny points of convergence between postmodernism and my reading of post-modernity drawn on global challenges. Now come the points of divergence. The first one is that global challenges reverse the postmodern *topos* which maintains that no substantive order exists in our world and no causal link can be clearly stated, thus dismissing any attempt to take a holistic look at the reality that surrounds us. They make us on the contrary recognize it as structured by all-encompassing forces and threats, which supersede the postmodern preference for the fragmented, flexible, vanishing. In one word, the old totality of the philosophy of history of Hegelian make is dead; nonetheless there are—alas, as they are indeed threatening—totality structures, against the "view of totality as a kaleidoscopic—momentary and contingent—outcome of interaction."[33] Moreover, global challenges also raise a question of practical philosophy, itself not the postmodernists' cup of tea: if the physical threats overshadowing us in a total dimension are man-made, should we not at least ask if they can be as well countered by human action? It is not said that they can, but if there is a chance they can we should then envisage collective and rational action, again something that enjoys bad fame among postmodernists, whose preferred politics according to Bauman are those of desire and fear. Later on this path will lead us to rediscuss the issue of reason in terms fairly different from postmodern *horror rationis*. Suffice it now to say that this postmodern attitude looks like high modernity's (eighteenth and nineteenth centuries) uncritical confidence in "Reason" merely put upside down, as more in general postmodernism is best seen as the mirror image of modernity with negative mark, that is, a position that does not really go beyond modernity.

This introduces us to the second major point of divergence: if we take the global challenges as a serious philosophical problem, we cannot share the postmodernist exclusive centrality of the Self for two reasons. It can be ruled out, or morally and even physically obliterated, by the "objective" structures and threats that do so much trim down the vast liberty of choice in which the late modern or postmodern Self seems to be at home, and almost reinstate the pre-modern primacy of fate (nuclear war, catastrophic climate change) over choice.[34] Then, the still open status of the question as to whether something can be done redirects our attention from a narcissist concentration on the individual Self towards issues of moral ties and political action.

Having said what keeps this book's view of what comes after modernity from postmodernism helps us to a further clarification. In its pages I have repeatedly used the formula "end of modernity *or* a major cleavage in modernity," which may suggest a close similarity between the two ways of periodization. This is not my view, as "end of modernity" implies that we are already in a new era, still to define, or at the end of history altogether, while in any case the failure of the modern project of redesigning our world in a reasonable way must be declared. On the contrary, a major cleavage in modernity would imply that modernity has become reflexive, criticizing itself and accepting limits to its essential drive for power over nature and society, but still pursuing its original project. The "or" I have put between the two outcomes means then that:

- The outcome of the present historical change is too early to call, we are still in the transition and only the next generations will be able to find out, hence let us keep the choice open; but this is admittedly just a rule of thumb of methodological prudence.
- As of now, both schools of thought contain some truth; the "end of modernity" school has the upper hand, and justifies my favorite formula "after modernity," because of the depth of the changes that came over us in a few decades: first nuclear weapons, then man-made climate change, lastly bioengineering, even if contemporary philosophy may have not yet developed full awareness of their implications and lingers over more traditional (social justice, community's disruption) or more urgent (multiculturalism, terrorism) topics. A new overarching theoretical effort to go to the roots of the coming *condition humaine* (*et du monde*, one must add under global warm-

ing) is not in sight. On the other hand, "a major cleavage in modernity" leads the attention on the necessity to rescue and redefine elements of the modern projects, as we shall soon see, if we want to come to terms with what is happening after modernity. It is perhaps useful to remind here that the modern project of redesigning human society had been undermined by total war and totalitarianism, in other words by Verdun, Auschwitz, and the GULAG archipelago, even before nuclear weapons and the other changes made their appearance.[35]

- At the very end, the choice for one formula or the other has relative value and may become a matter of linguistic convention, if the substantial elements are clearly spelt out. Also, what I am presenting here is just an assessment of how global challenges affect the course of modernity; they do so a lot, but this is not intended to replace a general theory of modernity and its periodization,[36] which should legitimately state whether or not it has come to an end. My self-imposed lack, as it were, of full jurisdiction is one more reason not to take irrevocably party in the dispute.

More important than the *querelle des modernes et des postmodernes* and the dispute about the end of modernity seem to be two substantive issues that are at the core of the cultural as well as political definition of modernity: what is changing in the modern (think of Machiavelli) relationship between ethics and politics under global challenges? This is going to be discussed in chapter seven, which shall be centered around politics. In this chapter we are now addressing the other seminal aspect of Western[37] modernity, the approach to science and technology.

3. THE ROOTS OF GLOBAL CHALLENGES IN MODERN SCIENCE AND TECHNOLOGY

Even in the popular imagination nuclear weapons are burdened on the scientists (as in the case of Kubrick and Sellers's *Dr. Strangelove*) and so is pollution on heavy industry's smokestacks or the individual cars' technology. Irresponsibility and "madness" of scientists, engineers, and drivers is charged with the dangers now coming into and from the atmosphere. Popular pictures of causal links and moral indictments find their match in intellectual patterns that have been developed as an answer to the tantalizing question: why has science gone adrift and unexpectedly generated more evil than benefits for human beings? We can recognize two such patterns:

1. Humankind has been unable to handle science and especially technology cautiously because if we glimpse a chance to further modify our natural and

social environment efficiently[38] and according to our plans and preferences, we cannot but do so. "Can implies ought."[39] This explanation relies on a philosophical anthropology to which human beings are weak towards other species, lack specific organs, and must produce for themselves increasingly powerful prostheses to come to terms with an increasingly complex world. This ability is thought to evolve into increasing power and lust for power, which becomes unable to perceive the environmental, social, and political limits to growth. This whole attitude is specific to nobody and rather assembled with elements derived also from Gehlen (man as *Mängelwesen* or defective being), Heidegger (the technical posture is telling about the nature of man and his way to approach the world), while "limits to growth" is indebted to Fred Hirsch's seminal contribution on its social limits. This position has the philosophical strength of referring our troubles with science and technology to deep roots, but in doing so it lacks historical specificity, in the sense that it does not explain why science and technology (and our troubles with them) rose to uppermost importance only as late as with the appearance of modernity. One step in this direction was made by the old Critical Theory[40] as it saw in the preponderant value given to self-preservation by the isolated individuals of the post-medieval, Cartesian/Hobbesian era the ground for accepting whatever device might prove instrumental to that absolute goal: this is the origin of the theory of instrumental reason, which in Frankfurt and elsewhere has played such a relevant role in the explanation of our modern troubles. Reason has perverted itself into an organ for devising the most appropriate means in order to attain goals that are not themselves under reason's control: this is how critics of modernity give an evaluative conclusion to the version of reason and rationality best analyzed in Max Weber's notion of formal rationality. On this path, rational and reasonable cannot but diverge or even collide with each other.[41]

Nuclear weapons fit perfectly this pattern, as they were invented as a supreme or absolute instrument of political or civilizational self-preservation, a goal not itself put to test by an encompassing human reason. Mass production for a consumerist market and a car for everybody with all their polluting effects did also look like rational means to obtain a standard of well-being, freedom, and even happiness which was critically examined neither in its logic nor in its consequences. Though, as Habermas in his critique of the old Critical Theory has argued, the notion of instrumental reason is too general and un-selective, because it takes under fire all types of instrumental rationalization in the several sectors of action. Modernity has not altogether been the demise of critical and humane assessment of the means-goals relationship, since it has been able for example to rationalize the economy as a subsystem of action in a way that does not require a permanent test on that relationship, which would downgrade its efficiency and the general well-being. Essential requirement is only that technical rationalization of our life-world alone does not expand into sectors, such as the cultural or biogenetic reproduction of society, that only make sense if they preserve their own, non-systemic logic. Now, the problems with global challenges are that they are not a threat to specifically "soft," communicative sectors of our world of acting, but to our civilized survival altogether, while deriving from sectors like security policy and economy that in Habermas' picture of modernity have been rationalized without excessive costs for humanity.[42]

2. The historical reference is defining for the other major pattern, which we can christen the "autonomous technology" pattern.[43] At the beginning were capitalism, interpreted by Marx as prevalence of the impersonal logic of labor force exploitation and profit-building, and the subsequent creation of abstract (i.e. non-communitarian) relationships of human beings among themselves and with their own products, dominated by the exchange-value. These considerations go beyond Marx and Engels into the *Kulturkritik* of the early twentieth century (Simmel, the young Lukács) and become some kind of conventional wisdom: our products tend to live a life of their own, far from the motifs, interests, and goals under which we generated them, and they rather impose over ourselves their own constraining logic, all dictated by "objective" or functional links and foreign to personal or moral considerations. It is the *Sachzwang*, the constraint that derives from the thing itself. Weapon systems and environmentally perverse side effects of our commodities can be seen as cases in point of this logic, although they were both unthought-of as the theory of the autonomization of our products was devised. A later subversion of this approach can be seen in Günther Anders' explanation of the atom bomb: modern technology has been so hugely successful that its tools have mesmerized human beings and misled them to replace goals by means, thus subverting the classical view, first formulated by Aristotle, on the relationship between *telos* and means to obtain it.[44] But, philosophical terminology notwithstanding, this must remain a shallow descriptive picture if it is not referred to an ontological or anthropological[45] structure. It converges indeed with what we noted in chapter three about the concern for security being mostly burdened in the nuclear age, eminently during the Cold War, on technology and its almost obsessive enhancement, instead on political measures. Now, at least in this case, the inversion of ends and means may find motivational support and anthropological explanation in an extended "unburdening" function (Gehlen's *Entlastung*) of technology: in front of a highly complex and threatening situation it looks like a relief to entrust machines instead of humans with the function of providing safety. This is not without an ambivalent historical dimension: modernity has always struggled to find a balance

between a romantic or Luddite technology-bashing and a gullible belief that technological solutions reducing or out-ruling the so-called human factor are best in all spheres of action.

On these grounds, "autonomous technology" is an expression that cannot be taken at face value, as it is sometime the case in conventional wisdom. As Marx used to say, commodities do not walk by themselves to the market; to be taken there to change owner, they need human actors, dominated though these be by the logic of the exchange value. Similarly, atoms of fissile material and the necessary metallic hardware do not assemble of their own force into nuclear warheads, nor does the polluting CO_2 come out of submarine oil wells and push into the atmosphere without being burnt by humans. "Autonomous technology" is best understood as a short for a twofold mental attitude of human actors: first, the technological idolatry that has found room in modernity supports the belief that, wherever a situation of social conflict or conquest of nature allows for being addressed by means of technology, this way is superior to any other, regardless of the externalities it generates now or in the future. Even before considering the edge it may confer upon its users in terms of power, it is chosen because it appears to be safer and better error-proof than solutions based on the "human factor." This is the first reason why technological advances, provided they are economically bearable, used to have leeway in modernity.[46] The second one is that technological revolution has been essential for capitalism, the first among the historical configurations of society (in Marx' words, capitalism as a *Gesellschaftsformation*) that not only shifted power from a class to an other, but subverted and reshaped the very mode of production, the technological and organizational methods of producing goods. Both in its broadest (like in Marx's view) and narrowest[47] manifestations the link of technology and economy is one more reason for not taking the autonomy of technology literally, but rather as a somehow misleading expression for the overwhelming fascination of technology on our minds.

It is however the first reason that seems to be more revealing of the deep motifs for this fascination. The argument I developed to highlight it needs qualification, because there is obviously no such actor as modernity or humankind, which we can hold causally and morally responsible for those orientations. Decisions are always taken by a particular group which identifies with the special interests of a social class, corporation, a nation, or an alliance, or has at least to put them into account, even if endeavors to take care of more general interests (say, the UN Security Council in its best moments, or a national government with a strong multilateralist or even universalistic concern) exist. But it is only in extreme conflict situations, like between Nazi Germany and the Western democracies at the dawn of the Second World War, that decisions such as to build what promises to be the "absolute weapon" are made under inescapable constraints. It is under these circumstances that a Schmittian friend-foe situation arises and makes striving for technological advantage a necessary choice. This lack of alternatives cannot however be granted with regard to the various stages of the US-USSR nuclear competition, in which—to put it bluntly—more diplomacy with less ideology and less technological idolatry could have found some space, had political and economic factors not been an obstacle. Much less inescapability, or none at all, can finally be seen in the steps that in most parts of the world have been taken in the direction of more greenhouse gases emissions, such as the absolute primacy given to individual mobility, the insistence on fossil-fuel based engines and poorly insulating construction techniques, requiring heavy heating and air conditioning. In all these cases, beyond the various decisions made in one or the other country, there really seems to be something like a widespread cultural trend that makes not just the ruling elites but the majority of the "citizens of the market" (*Marktbürger*) prone to hail whatever technological novelty irrespective of its costs and consequences. This is perhaps the core reality behind the "autonomous technology" formula.

But the proneness to technology as a first choice is not simply the eruption of some anti-humanist idolatry for the machine as the new Golden Calf. We should still consider what its presupposition and its nearest cause are. Technological advance would not be, and for millennia was not available as a choice or an idol as long as modern science did not make it possible, since it changed (mathematized) our approach to nature, and introduced the experimental method, thus penetrating the "secrets of nature" like no culture before; it also introduced the new ideal of pushing back natural necessity and using the new, however lopsided, knowledge of natural laws as the tool to reshape nature and improve the human condition. Without these practical and philosophical scenarios the notion of technology becoming autonomous would have never had a chance. This presupposition is to some extent still alive: with or without the Baconian purpose of general human progress, the attitude to scientific research and discovery of this and other worlds is still recognized in most of the leading cultures as fundamental and positive, something that cannot be renounced,[48] it is not going to be the "qualitative" or esoteric approach to nature preached by new sects and religions to put an end to this essential feature of modernity, of which we are still part insofar.

If this was and still is the presupposition of the proneness to technological idolatry, there is no direct way from the presupposition to this outcome. Science as advance of knowledge accompanied by the "joy of insight"[49] must be conceptually kept from technology, instead of merging them together, as bad sociology and vulgar Marxism used to do. The effective historical link between them in our modernity (can there be an other one?) has its cause in the enormous success of technology driven by science in changing our life for good or evil; a success achieved first of all in most of the world by rapidly extending its duration for the individual beyond limits unimaginable still a century ago, making illness less or later threatening and thus postponing at least the fear of death which is at the root of so many mental and social devices of our kind. It is not just the amount of change, but its incessant, revolutionary progress and the speed of this process that have surpassed the human ability to keep pace with it, to generate a cultural change that matches the change in the material supports of our daily life and the behavior patterns they impose over the humans. Not only philosophy, religion, and the arts have been mostly and for a long time unable[50] to live up to the new settings of our material and psychological life and to avoid simple embarrassment or reactionary postures such as romantic anti-capitalism of leftist and rightist make; it is the political cultures of both elites and ordinary people that have failed to grasp that something fundamental was changing and politics needed to be rethought.[51]

Our difficulty to mentally catch up with the triumphal march of technology has been interpreted by Anders as a variation on the "Promethean gap" (cf. chapter four, § 4), that is, as a gap between scientific knowledge and capacity to conceive of what is going on (*Wissen* vs. *Begreifen*), between mass killing and ability to repent, between the human being while producing and while feeling; to bridge the gap we need "moral phantasy" as well as a broader capacity to feel,[52] a new chapter in the history of sentiments. After Anders wrote this in 1954, something close to his requests has emerged in the anti-nuclear war movements of the fifties and the eighties[53] as well as among the people with a "green" mind in movements, parties, and science communities. Something like a cultural revolution in the culture has started and is now having some significant impact on policy, both nuclear and environmental.

This is an example among others (not many however) that after the near collapse of modern rationality in politics and in the use of science, signs of a new rationality are popping up, although I would prefer to use the adjective rational or, rather, reasonable[54] without resorting to a substantive as burdened as rationality. New rationality, if any, is different from the modern one: it does not have pretensions to reshape the world, just to make it less inhospitable, it is able to criticize itself and to respect its own limits, it does not deny links to emotionality and imagination, as it is essential in our attitude towards our posterity. In this frame science remains an irreplaceable component of a reasonable approach to the world, once it has been freed of all pretensions to be the paramount paradigm of rationality and progress, and after the responsibility for man-made threats has been redistributed among science, technology, and policy-making, as I have been trying to do; to mention just an example, the chapter on climate change has provided new evidence in this sense.

May a cultural revolution towards a different rationality have started or not, it would be in any case like a mill that grinds cereals as slowly as to let everybody die of starvation before wheat is ready and bread is baked. This is the image used some seventy-five years ago by Sigmund Freud to express his perplexity about the chance of cultural renewal, in particular new libidinal ties among nations to avoid a renewed total war (cf. Freud 1932); it took unfortunately only a few years, until 1939, to prove *ad abundantiam* that perplexity true. We cannot be happy once we have rightly recognized that more moral and emotional imagination would help the people to grasp the depth of the threats and to make actual the obligation towards the vulnerable future generations, whose expected suffering remains otherwise the reasonable but purely theoretical notion (vulnerability in its link with responsibility) that we have seen in chapter five. We now know that our modern troubles with the technological transformation of science are less a matter of economic constrains or autonomous self-development and rather a problem of our mind, of the way we can master new chances and developments, and rethink the possible course of action we can select under unprecedented circumstances (for example, the chance to build the ultimate weapon, or the rising forecast of dramatic climate change). Particularly in this last case, it is crucial how our collective expectations are shaped in terms of what we regard as desirable, permitted, forbidden, advantageous, or disadvantageous for the community of humankind and/or for our particular community. Now, the dimension of interaction in which they are collectively shaped is called in political and social philosophy an *institution,* that is, a set of rules that is persistent over time, interconnected without self-defeating contradictions and major loopholes, while covering a whole dimension of cultural, political, or social interaction; last not least, this set of formal and informal, statute and customary, legal and conventional rules must make sense to the actors according to their universe of possible meanings, otherwise they could not be enlivened.[55] To make this even more clear, we are not going to discuss here the rightness of normative rules, or whether or not we ought to follow them; we just look at institutions, may they be social such as "mutual respect" or political such as "the state," to find out how far they are effectively or presumptively up to global challenges. To appeal to the "responsibility of scientists" or to tout the "precautionary principle" may be significant for individual morality, but lacks any true and stable influence on where technology leads us if it is not supported by a political authority capable to set and to enforce standards and rules.

Of the different institutions that may be relevant to how we and our predecessors and posterity come to terms with the two global challenges or fail, we cannot possibly discuss the entire range. As we are eager to get some bread still before the old mill of cultural change has finished its work, we shall focus on those institutions in which cultural, moral, or religious change should be brought to bear and help generate decisions that may effectively give some relief from those threats. Decision-making on collective goods implies power relationships supported by force and questioned in terms of legitimacy. This is why *political* institutions are one of the main subjects of our seventh and last chapter. A political institution, the territorial centralized state, was the device men invented at the outset of modernity in the attempt to come to terms with the new challenges, as we have seen in chapter three, § 3. We are now going to ask about the role of political institutions in front of the new post-Leviathanic challenges.

NOTES

1. The title of this chapter echoes that of a work by my teacher and friend at Heidelberg in the sixties, Karl Löwith, whose *Meaning in History* was written during the American exile (Löwith 1964).

2. Not necessarily one life plan for the whole of our life, as Rawls may lead the reader to think of it (cf. Rawls 1971, § 63) but possibly—especially in postmodern times—several life plans in different ages, which confronts us with the task of re-adapting the underlying narrative while modifying or replacing former structures of meaning, something we can do as long as we can legitimize the shift in our symbolic universe. In criticizing Rawls I do not go as far as Larmore 1999, who discards the very idea of a life plan.

3. The link between meaning and value as a way to transcend the limits of individual life has been discussed by Nozick 1989, § 15.

4. In the *Odyssey* Achilles, talking to Ulysses in the kingdom of shades, sticks to the superiority of whatever kind of life over death, whereas the chorus in *Oedipus at Colonos* 1225 maintains that "never to be born is far best," and so do Byron and Schopenhauer; for these and other examples cf. Singer 1992, 137.

5. See Hobbes 1651, part I, chapter 6, 118–30.

6. Thomas Nagel has a more pervasive notion of absurdity, which comes—as he put it—from the clash between the seriousness of life and the possibility of regarding everything as arbitrary and doubtful, see Nagel 1987.

7. Thus the symbolic universe links men with their predecessors and their successors in a meaningful totality, serving to transcend the finitude of individual existence and bestowing meaning upon the individual's death," Berger and Luckmann 1967, 103.

8. Remember however the deep impact of the catastrophic Lisbon earthquake on an intellectual hero of modernity like Voltaire.

9. The links between modern science and philosophy, particularly Cartesian, and sovereign statehood are examined in Swazo 2002 from a Heideggerian approach.

10. On this twin process cf. Habermas 1984–87, chapter viii, § 1. On the difficulty to find meaning for our collective life under globalization see Laïdi 2001.

11. This point was made by Bobbio as early as writing in the sixties, see Bobbio 1979.

12. The intergenerational extension of our life plans, one could say.

13. This is the meta-ethical aspect of the problem of the meta-imperative discussed in chapter five.

14. The Greek name of the mind.

15. On the role of autobiography or résumé (*Lebenslauf*) in the production of meaning see Schütz and Luckmann 2003, chapter 2, § 6.

16. Cf. Toulmin 2001, 209, on the "rationalist" program of the new philosophy of nature from Descartes over Newton to Kant.

17. In this frame of reference it would be interesting to look back into episodes such as the mass-suicide of the American sect of Reverend Jones in Jamaica and the tragic end of the Branch-Davidian sect in Waco, Texas, in 1993.

18. On the destabilizing effect of work conditions under globalization on the character's build-up see Sennett 1998.

19. Weart 1988. An empirical social-psychological inquiry into German fears at the top of the Euromissiles crisis in the early eighties is found in Volmerg et al. 1983.

20. This wording is taken from the title of Martin Wangh's contribution to the volume edited by Levine et al. 1988, still the best survey on the subject. See also Nedelmann 1985.

21. See Lifton and Falk 1982 and the former studies by Lifton mentioned there. Similar studies on the impact of global environmental threats are so far not known to this author.

22. It is worthwhile to note that the Bush administration, as it wanted to create a psychological climax in favor of waging war on Iraq, resorted to nuclear fear, alleging an actual or impendent nuclear capability of the Iraqi regime.

23. Or *Gattungsethik*, see Habermas 2003.

24. *Vorwelt* and *Nachwelt*, to put it with Alfred Schütz, see Schütz and Luckmann 2003.

25. They may suffer from the consequences of global warming, but they would not be able to see it as a worldwide phenomenon with roots in human activities that could be changed.

26. The inflationary and often fuzzy use of the words global/globalization is not helpful in this sense.

27. The surveys quoted in chapter four, note 31.

28. Cf. Lyotard 1984 and, as a summary of his abundant writing on this subject, Bauman 1999.

29. Philosophy of history is in this sense a narrative that sees in the chain of historical events a plan set up by God or Reason (Hegel, in some sense still Marx) or Nature (Kant, in a moderate version; positivism and positivistic Marxism) and a progress of humanity towards the goals or models prescribed by them.

30. Or, more comprehensively, of the "modern project" (*Projekt der Moderne*), as Habermas put it.

31. Horkheimer and Adorno 1947. Written almost entirely in 1944, it was first published in 1947.

32. This is not to deny the importance of Marx' lucid analysis of the link between profit creation and damages inflicted to the environment, or—as he put it in the first book of *Capital*, chapter 13—the earth.

33. Bauman 1999, 86. Oddly enough, this view resonates with the unfettered "processualism" of the early Lukács' Hegelian Marxism in *History and Class Consciousness*, in which only processes, neither structures nor institutions, are recognized as basic ontological structures.

34. For this point see Berger 1977, 77.

35. In the three years between the first draft and the publication of *Dialectic of Enlightenment* (cf. footnote 31) Horkheimer and Adorno were able to grasp the significance of Auschwitz, adding a section to their chapter on antisemitism, but did not make any mention of Hiroshima and Nagasaki—nor did they in the decades thereafter recognize the philosophical significance of the atom bomb.

36. For example, Scott Lash sees a first modernity run by the logic of technology, a second one centered around the logic of culture, followed by a third dominated by technological culture, cf. Lash 1999. From Bruno Latour's *boite à surprise* (1993) the (for us scholars of modernity) shaming news pops up that we have not yet been modern, which brings us back to the point that even in philosophy much is matter of linguistic conventions and the puns we can make upon them.

37. Science at an advanced stage was present also in the Chinese and Arab civilization, but the way it was developed into revolutionary technology and put at the center of civilization was specific of the European West.

38. That is interposing between us and the modifiable object not just whatever means or *Gestell* (Heidegger 1977), but means whose costs are decreasing towards earlier solutions.

39. Lenk 1982, 35–36 speaks of a mechanism that makes normatively obligatory whatever is technically possible.

40. Reference is made to Horkheimer's essays of 1941–42 on reason and self-preservation (reprinted in Ebeling 1976, 41–75, a volume which is still a seminal work on the notion of self-preservation) and the authoritarian state (Horkheimer 1982).

41. For this terminology see Toulmin 2001. I am aware that rational-reasonable does not match properly the German pair *rational-vernünftig*, which I am rather having in mind in these reflections. A broader discussion of this seminal pair, which should include Rawls 1999, is not possible here.

42. This discussion looks back at his systematic theory of rationality and modernity, see Habermas 1984–87.

43. This is the title of book that has best highlighted this approach, see Winner 1977.

44. This notion of ends-means inversion is found in the first and still most penetrating philosophical essay (*Über die Bombe und die Wurzeln unserer Apokalypse-*

Blindheit) on the atomic bomb, written by Anders (whose original name was Stern, a former assistant of Heidegger) in 1954, see Anders 1956, 233–324. Curiously enough, no English translation seems to exist.

45. Wherever not further specified, philosophical anthropology is meant, with links to, but not identical with the German *philosophische Anthropologie* of the years between the wars.

46. I am using the past tense not in the belief this attitude has disappeared, while it has indeed become matter of struggle and is no longer uncontested, at least in the West.

47. By narrow manifestation I mean the sheer dependency of technological (and even scientific) advances on military spending, on the "military-industrial complex" (Dwight Eisenhower), or the gridlock imposed over the introduction of cleaner engines by the car and oil industry.

48. In its various shapes and images, for Srđan Lelas modern science is characterized by three ethical standards: free access to the results, predominance of the critical attitude over political negotiation, and (epistemic) respect for the object (I would deem the latter should come first), see Lelas 2000.

49. See the quote from Weisskopf 1991 in chapter three.

50. With some exception obviously: just to give scattered examples, Futurism in the arts, Hans Jonas and Lewis Mumford in philosophy.

51. Only a few hours or minutes before the catastrophe that could have resulted of the Cuban missile crisis did the Soviet Secretary general and primarily the U.S. President show that they were aware of the jump made by technology-based destructiveness; they would have otherwise fallen back into the "guns of August" syndrome and pulled the trigger (I am hinting at the title of the book by Barbara Tuchman that helped JFK not to slide into the same patterns as at the outset of the Great War). As for the environment, this leap of awareness is still incomplete, to say the least.

52. *Erweitertes Fassungsvermögen des Fühlens*, Anders 1956, 315.

53. In the fifties the Communist parties played a major role, while the antinuclearism of the eighties was much more anti-American than anti-Soviet, although the last arms race of the Cold War was started by the USSR. But the new sensitivity that motivated the participants was a cultural novelty regardless of political instrumentalization.

54. Cf. Rengger's recent appeal not to throw "the reasonable baby out with the rationalist backwater" (2005, 327). This language usage is indebted to Toulmin's *Cosmopolis* (1990).

55. The notion of institution has been already introduced in chapter three, § 4.

Chapter Seven

The Politics of Global Challenges

After two philosophical chapters dedicated to the consequences of the two global challenges for our normative morality as well as the meaning structures that make it possible to live our lives, we are finally coming to the concluding issue: what does this all mean for Leviathan, that is, for modern, state-centered politics? Or must ethical and political aspects of the global threats remain unrelated, as at least realism, a major strain in modern thought, suggests? These are epochal questions, they require a fair deal of theoretical engagement on where the crisis of political modernity may lead. We will see in § 1 that neither separation nor merging of politics and morality are adequate solutions to puzzling new questions about their relationship. Provided with this sense of complexity, I will review in §§ 2 and 3 the various grand strategies that are proposed in order to meet the new challenges, more explicitly the nuclear one: realists suggest more nuclear anarchy as a promise of stability, while for radical cosmopolitans a world state could eradicate anarchy. Based on a skeptical assessment of these proposals, in § 4 I shall then argue that, while old Leviathan is unable to live up to the new challenges, on the other hand we can see some modest signs of change in the political and legal culture; they could introduce an attitude more appropriate to the scope of what we are confronted with. In the last section some policy proposals, including the abolition of nuclear weapons, are discussed, while the primary thesis remains that neither politics nor philosophy can now be credited with providing us safe havens as solid and protecting as the modern Leviathan used to be.

1. BEYOND REALISM AND NORMATIVISM

Which consequences can we draw from the first two chapters of this Part Two? We have learnt that the two-track (obligation and meaning, normative and meta-ethical arguments) reflection on the global challenges converge in telling us, humankind, to do the best we can to prevent their appalling consequences from coming into being.

To give reality to this obligation is a task for *politics*. We may deem a cultural transformation that leads us to value peace and environment above anything else to be a decisive factor of prevention, but this process must in any case end up in politically negotiated acts of disarmament and emission reduction. On another path, we may regard technical innovation (a flawless anti-missile shield, a new cheap and clean source of energy, speculations that these may be) as the key to the solution of the problems with the present global threats; but also in this case the innovations are to be implemented by taking care of the political context, if they are not to yield perverse fruits such as the instability caused by a unilateral technical advance, which enhances only one party's power. However, to think of politics as the (low or shallow) activity bound to realize morality's higher views into humankind's future would be a naïve picture of how politics (and even more, policies) and morals interrelate. This simplifying background picture is the reason why so much otherwise valuable ethical literature on global evils is unable to live up to the earnestness of the tragedies it endeavors to highlight. It is not simply that it underevaluates the awkward task to bring politics closer to the standards of morality; even when it does not do so, it conceives of the emerging difficulties as a resilience of politics against moral pressure or a transient deficit of morality in subduing politics under itself. This misses the point that morality and politics are neither fully separable, but rather bound to live in some perpetual tension between them; nor are they two levels of action of a vertical entity called "human being," one of which (in most, but not all doctrines, morality) is bound to prevail over the other. They are rather two realms of action on a roughly horizontal surface, each with its own logic, often striving to ignore or, when they come to overlap, to overwhelm each other, relatively independent but never fully isolated from each other.

This is admittedly nothing more than *one* interpretation of the relationship of politics and morality *in modernity*. Let us now look at the two leading doctrines, realism and normativism,[1] which are concerned with that relationship. In the case of the former, the very presuppositions of its disjunction of (individual, normative) morality and politics have waned. Political decisions regarding global challenges are not a problem of simply shifting the balance in the unequal and conflict-laden distribution of divisible goods such as land, seaports, glory, or prestige. Security against nuclear annihilation or protection from the oncogenic effects of the ozone-layer depletion are not fit to be the stakes of a zero-sum game, which has been the ground pattern of politics from the Trojan war till the Cold War, and remains obviously fundamental wherever politics means adversary distribution of scarce and indivisible resources. But politics implies since 1945 a second substantial pattern, in which the survival of humankind and/or civilization is the prize to win or to miss. Near the classical goal of the *salus reipublicae* the new, somehow competing need to care for the *salus humani generis* has come up; the former presupposes that the latter is not a problem, which is no longer true, and the inability to reinvent the former in accordance with the latter, in other words, the permanence of the security dilemma, can doom humankind as well as the *respublicae* or Leviathans. Under these circumstances Machiavelli's claim that the *salus reipublicae* has a moral stand no lesser than that of individual morality must collapse, because the state can no longer be credited with the ability to guarantee adequate conditions to human life, let alone the flourishing of the citizens' republican virtues. In this light, the recent revival of republicanism under Machiavelli's aegis, based as it is on the doctrinarian attitude not to take the present state of the world, looks bizarre not only with respect to the global challenges, but also to the anthropological and social transformations of the citizens, particularly in Western democracies under globalization.

This also means that the span of future influenced in a dramatic, even lethal manner by political decisions is incomparably larger than the life span of Machiavelli's *principe*, of his (or her, in case of a *principessa*) children and grandchildren, or their subjects.[2] At the same time, technology has lost its moral neutrality, and so has the political decision-making that affects its evolution. Questions concerning our responsibility for future generations become unavoidable, and are not as optional or fictitious as the care for the country's future glory used to be, linked as they are to physical and verifiable evidence of impending evils. Nothing similar existed as our traditional notion of politics was shaped at the beginning of the modern age. The background worldview of political realism, that is, its notions of time and history, individual and collective fate, has been losing its foundation in our experience. This must be taken into account before we go over to review realism with regard to the political solutions it suggests, which will happen in § 3.

While the distinction of politics and morality introduced by *modern realism* must be rethought, but cannot be simply discarded, the radical view of an absorption of morality into politics advanced by two icons of realism such as Hobbes and Hegel has lost ground. As we have noted in chapter six, the nuclear and now the environmental (climate change) threat have ended the job, initiated by Marx, Nietzsche, Freud, and the First World War, of stripping Hegel's faith in a rational destiny of humankind guided by the *logos* of any residual credibility. Hence the fall of the Hegelian pretension of historical reality to be in all circumstances superior to the claims of normative, particularly deontological morality. But Hobbes' belief that all evils of human nature can be met only by the political subjugation of the individuals to the sovereign, with no margin left for conscientious imperatives,[3] has not fared much better, since his principle of the commonwealth as the sole and supreme guarantor of peace and security has been disproved by the nuclearization of the Leviathans, which promised to sustain order and peace among human beings and can now bring about their annihilation.

However, the other pole of modern thinking about morality and politics, the Kantian paradigm,[4] is of no greater help than realism in addressing the human condition defined by global challenges. Gone is not only his thick notion of Reason as the force which tells us to subordinate politics to binding moral imperatives, but also the guarantee that, in doing so, we will not simply bow to moral law, but also achieve perpetual peace sooner and more firmly than resorting to political prudence. Kant himself was seemingly not content with the pure moral statement that human beings ought to respect rational morality, and tried to outline a political argument (effectiveness of perpetual peace as a result of "moral politics") to make his proposal acceptable with regard to its consequences and not only to its principled rightness. But the foundation of this "guarantee of perpetual peace" on a presumptive "mechanism of nature" generating harmony (*Eintracht*) out of conflict (*Zwietracht*) was the sign of his difficulties rather than a winning argument.[5] On the whole, Kant's political proposals aiming at creating conditions for perpetual peace ("republican" or liberal-democratic domestic change, inter-state federalism, cosmopolitan law binding states to acknowledge certain rights to other states' citizens) have put his fame as leading representative of *political idealism or normativism* on a firmer and longer lasting base than his view of politics as a realm merged with and subdued to deontological morality. This argument might receive further evidence from reflections on two little known passages in Kant's secondary writings.

Insisting on the subordination of politics to morality and morality-based law, he supports the old dictum "fiat iustitia, pereat mundus" adding the clause "whatever the physical consequences may be [. . .] The world will certainly not come to an end if there are fewer bad men (*Menschen*)" (Kant 1795, 124).

In the nuclear age, no morality can sustain its claim regardless of the physical consequences, because the world may very well collapse while a nuclear-armed morality pursues those who are regarded as *böse Menschen*. On the contrary, any morality and morality-based policy must acknowledge and implement the meta-imperative aimed at the survival of humankind that I have made the case for in chapter five. In the other passage, from the already mentioned article of 1794 on *The End of All Things*, Kant concedes, though with regret, that in order to drive the human mind to moral conversion the belief in the value of virtue is less effective than the emotion driven by the picture of the world coming to an end under appalling conditions (Kant 1794, 225).

On a more general plane, it cannot be said that the end of political modernity and the upcoming of global challenges have discredited a normative (deontological) approach to politics because two hundred years after Kant we are struggling with the scourge of a much more terrible war than Kant could dream of; the failure of a principle to bring about change in a limited time is not an argument against it. In the article *On the Common Saying: 'This May Be True in Theory, but it Does not Apply in Practice'* (Kant 1971, 89) Kant himself points out that the rule "if it has failed so far, it will always fail" can be used neither in technical questions such as the possibility to travel by aerostatic balloon nor *a fortiori* in moral ones such as the moral betterment of humankind. It is rather the structure of the problems we have to deal with that makes pure normativism no longer viable, at least as a guidance to political action. Its basic rule "act justly regardless of the consequences" becomes self-defeating if only the slightest possibility exists that the consequences bring about more evil (human sufferance, unmoral, and uncivilized conditions of life)[6] than that addressed by our moral agency. This possibility has always existed, but its scope has been enormously widened by our unbounded technological manipulation of nature. If technology has thus lost the moral neutrality enjoyed in modernity, the reverse holds true as well: *morality is no longer innocent* of its technical and social effects. To assess them however is competence of politics and science, not individual morality. In case of conflict between these forms of knowledge, neither can claim to be superior by default to the others; the conflicting indications for action must weighed against each other argumentatively, and there is no guarantee for the conflict to be completely resolved. Moral arrogance, whether deontological or utilitarian, is no less disturbing than political cynicism in front of the often contradictory complexity of human life in both its individual and social spheres.[7] When morality becomes moralism, it reveals the inability to grasp and to possibly balance the several moments of this complexity, and in its self-righteous simplifications also damages the chance of a reasonable normative approach to collective threats and challenges.

As a conclusion, the vicissitudes of modernity have led Kantian normativism, whether moral or legal, to come to terms with the category of *consequence*, originally a distinctive feature[8] of normativism counterpart, consequentialism, which includes utilitarianism. Far from giving consequentialism victory over deontologism, I want only to stress that the traditional self-sufficiency of modern theories of morality is losing ground, and that we must look for new road maps. In this book I am not going to do so in general meta-ethical terms, not just because this would let the book implode, but primarily because I deem a general redefinition less fruitful and also less feasible as of now. We need to learn more about politics and moral problems after modernity before we can sketch such a definition. Global challenges are not the only novelty they have to go through before they can find a new balance; globalization in general has generated other novelties, for example the globalization of human rights on the one hand and, on the other, of terror by non-state actors, a fact that challenges the modern version of *ius in bello*. In any case, a substantive approach, which looks into the variable interplay of politics and morality by studying the concrete issues they have to deal with, seems to be more productive and better up to postmodern times than pursuing a systematic redefinition of their relationship.

Only one element of a general redefinition needs to be mentioned explicitly. It is easy to understand that pointing at consequences as control instance on moral normativity means stressing *responsibility* as a category relevant to any moral theory that wants to have a saying in politics. This is obviously well known since Max Weber's considerations in *The Profession and Vocation of Politics* (1918) on ethics of responsibility and ethics of intention being both necessary to politicians and statesmen. But new aspects have come up since then, among them the responsibility we have new reasons to feel against future generations, as we have seen in chapters five and six. On the side of politics, this means to play this responsibility against acting just in favor of the present generations' immediate interests. But on the side of morals, responsibility has to contain an uncontrolled agency that is guided either by much-promising projects for the future of society,[9] or a moral rigorist attitude that wants to immediately translate moral norms for the individual into public policies, or lastly a moral predication unconcerned as to whether its allegedly just doctrines are really promoting a just behavior among the public. In short, responsibility is a category that binds together morality and politics, none of which has a monopoly on it. As a purely moral category, unaware of the specific logic of politics as well as technology, *responsibility is an empty tool.*[10] On the other hand, living up to one's own responsibility in politics cannot be done nowadays in the short-sighted manner whose flavor is attached to the formula "a responsible politics," as the new problem of a responsibility towards future generation can no longer be skipped.

It is time to examine the concrete configurations of political and moral, but also anthropological and psychological elements that reveal themselves while we go through the main strategies on how to cope with the global challenges.

2. NEITHER EMPIRE NOR COSMOPOLIS

Until now political grand strategies have been devised as to how to tackle the nuclear weapons problem, while man-made climate change has not unleashed a debate as to whether world politics needs to be reshaped for its sake. At least in the present perception, this problem does not look as dramatic and acute as nuclear war or peace, and this is why intergovernmental cooperation has been seen as sufficient to come to terms with it. I have made in chapter four the case for global warming being philosophically and morally as serious a problem as nuclear weapons, and this belief concurs in setting it on the agenda of a politics for the future generations, connected to interests and identity of the present generations. This raises the preliminary problem if a "politics for the (far) future" can exist, which a realist theorist as well as an empirical political scientist would probably incline to deny, insisting that "all politics is present,"[11] that is, has to stick to the present actors' short-range preferences, and that *a politics for the future* (a short for what we have mentioned in the previous sentence) may never become future politics. Now, along with other structural aspects of the role of time in politics, the time gap between the interests and beliefs of the present decision-makers and those of the future people at the receiving end of the decisions is a major problem; but a problem is a problem, not a taboo that should prevent us from imagining and discussing what a politics projected into the future may look like. Further, we are now going to discuss this question not just because normative reasons from chapter five tell us to do so. This matters, but in chapter six we have already seen other, non-normative cultural motives for introducing the perspective of even the far future as an element of political action; other hints in the same direction, for example from legislation and jurisprudence, will be given in § 4 of this chapter. On the other hand, a preoccupation with the future, not necessarily only the next future, is present in the realist literature on how best to redress nuclear anarchy and is not the monopoly of idealists. To conclude, it is legitimate to extend to global warming the reflections about future regimes that have been originally designed to cope with nuclear weapons, provided this is done respecting each item's configuration. With comparison to the present state of the art, it is a widely fictitious extension, as in the literature on global warming hardly any alternative to intergovernmental agreements is presented, and the problem is rarely talked about in nearly the same dramatic terms as about nuclear weapons.

In any case, for both global challenges the *structure of the problem* is essentially this: if these threats affect all men and women and all their communities, and can (because of their technical nature) be addressed with some chance of success only if all of them agree (even unconsciously or tacitly, for that matter) on the same solution, how can this be reconciled with the persisting, perhaps non-eliminable division of humankind in separate, independent actors with no superior power (*tertius super partes*) capable to restrain them, that is, with anarchy? In this context "technical nature" refers to more than just technology; it means that the destructiveness of nuclear weapons is not neutralized if everybody disarms except one bad guy who can then blackmail or destroy all of the good guys, or that global warming is not or not sufficiently slowed down if everybody cuts greenhouse gases emissions except one or two big countries that would then spoil the good guys' efforts, also with respect to the impact of their emissions on the existing inertia effect. The reader should also note that this formulation of the problem does not imply any particular instrument (world state, or imperial world power, or spiritual conversion to peace and solidarity by all citizens and statesmen) that is thought to best lead to the desired outcome: a uniform behavior by all decisive actors, which would eliminate the disadvantages of anarchy.

Fundamentally, three divergent solutions to our problem are available in the theoretical and policy debate:

1. exiting from anarchy by surrendering all power to a Superleviathan, that is, a benevolent world state (under democratic and environment-friendly, cosmopolitical premises) or a less benevolent, but still protecting empire; or at least surrendering all power in nuclear and emissions matters to central sanctioning authorities.
2. enhancing anarchy by endowing all major players with nuclear weapons and/or anti-missile shields, and renouncing any worldwide regulation on emissions, in the expectation that peace and emissions reduction are brought about by the internal mechanisms of political and economic anarchy. They are in the first case a lasting omnilateral deterrence based on a nuclear balance of power, in the second a mix of market mechanisms such as emission trading based on local or national regulations, or market-driven technological innovation leading to emission reduction (see pp. 200–01).
3. regulating anarchy, bringing more and more elements of mutual respect and cooperation into it, and above all promoting the redefinition of the actors' selfish interests in a more inclusive way and on the basis of an emerging shared identity as humankind (see § 4).[12]

I shall argue that 1 and particularly 2 are analytically unlikely or unviable and normatively undesirable solutions, while 3 opens new ways to inquiry and policy-making, although it cannot free us from threats that will stay with us forever, thus making humankind's future pretty different from its past. We will see that, beyond what may occur to real things (nukes, emissions) according to the different approaches, relevant and intellectually stimulating are in any case the conceptual innovations and the changes of mentality that result from projecting global challenges into the future.

Let us now discuss individual aspects of the proposed strategies.

On 1 The *empire* is too unlikely as to leave much room to the question if it would be absolutely undesirable. I think not, as it would at least for a time give protection from the nuclear threat; while, admittedly, only a very fatherly empire would also order the reduction of emissions. An empire is not a well-received form of government in our times and perhaps also in the future, and, worse, a planetary empire evokes Kant's fear of a "soulless" despotic regime.[13] But under it humankind, no longer divided in contentious nuclear Leviathans, would survive, and empires can change in some cases to the best, as seen not far ago in Gorbacëv's Soviet Union. However, a worldwide empire would be perceived as more unbearable than other types of world state, thus unleashing rebellion, becoming more oppressive, and finally collapsing into follow-up states, most of them with nuclear weapons, which would reproduce the dangerous anarchy existing before the empire rose. Once again, think of the Soviet Union, with three follow-up states (Russia, Ukraine, and Kazakhstan) creating an unprecedented type of proliferation; although the swift denuclearization of the two less powerful states has been a promising case not only of counter-proliferation, but of a changed mentality as well.[14]

But we do not need to debate whether good or evil would prevail in the case of the worldwide empire, because it is hard to believe that something like this would materialize in the next decades, if not centuries. It is not only that the only credible candidate, the United States, has proven incapable to establish its full rule over the present world and even to prevent attacks on its own territory, premises from which a development towards a planetary ruler is unreasonable to imagine. This is obviously not to deny that in some corner of American politics and ideology imperial temptations exist, nor that the United States has overwhelming influence in world affairs,[15] nor that its influence has been recently managed with a unipolar approach. But influence as well as military intervention is still far away from the direct and unified command on political and military power worldwide. Regardless of how well or poorly America's chances to become a true empire are set, it is the very structure of the existing nuclear multipolarity that makes imperial developments highly unlikely. It is not credible that the other nuclear actors would give up their arsenals in favor of the United States, while on the contrary the real trend is towards proliferation as the ultimate tool for resisting absorption under the "Great Satan" (Iran) or, plainly, the rules of the international community (North Korea). In this scenario, only victory over all other owners of nuclear weapons and their subsequent concentration into one hand would create the peace-preserving empire, but this would then be scarcely an empire and rather Schell's "republic of grass and insects."[16] A self-defeating process of founding an empire indeed. This foundation was, if ever, possible only once in modern history, precisely in the four years (1945–49) in which the United States was the only nuclear power. Scientists such as Albert Einstein, Robert Oppenheimer, and Edward Teller claimed that only by uniting in a world government the peoples could avoid internecine nuclear wars, while others such as John von Neumann and Bertrand Russell went so far as to advocate a preventive war on Russia, in case this country did not adhere to the world state; as the like-minded U.S. Secretary of the Navy Francis Matthews put it still in 1950, the United States was encouraged to become "the First Aggressor for Peace."[17]

Does the perspective of a *democratic world state* deserve more credibility in analytical terms and more support on the normative plane? Instead of an unlikely and nasty imperial Superleviathan, should we opt for a democratic and lawful one? At the first sight, yes. The forceful idea behind all variations of the cosmopolitical proposal is made of:

a. the assumption that the amount of problems and troubles of worldwide scope has become overwhelming and can only be addressed by the united humankind, while the common issues are on the verge to supersede the particular interests of nations and groups;

b. the belief that, democracy being widespread among nations, it can also shape the worldwide commonwealth to come, as it remains the only political regime that can harmonize with the *free* will of uniting in order to improve everybody's security and life chances.

This is obviously a simplified standard view of cosmopolitanism, an approach which goes back at least three hundred years in Europe's and America's intellectual history and presents a wide array of different versions.[18] I am not going to examine them in any detail and want instead to clarify that the settings for my discussion of cosmopolitanism are given by the contractarian, especially Hobbesian account of how the state is originated. It is the intolerability of the threat for life and limb in the pre-political state of nature that pushes the human individuals to uniting among themselves, while giving up their unlimited freedom or Hobbes' *ius in omnia* (rights on everything), and submitting to a supreme authority. On the other hand, the second-level actors, the states, are not subject to the necessity to unite, as we have already seen in this book.[19] The question is now if the threats resulting from nuclear weapons and man-made climate change are *changing the Hobbesian situation* and creating such a necessity, or in any case enough motivation to do something that goes beyond taking care of the citizens' life by acting only as an individual and particular actor, or in other words as the Westphalian separate, sovereign, and selfish state. Theoretically, this actor could be replaced either by a unified actor, even if not necessarily a world state, or by a network of redesigned states, which we can think of as less separate/individual and, what is more, less selfish/particular than the modern state we know. We will explore both these possibilities, but first of all let us recall the reason why I have set narrow limits to the notion of global challenge and, as a consequence, acknowledged only two out of the many possible candidates; which is at the same time the *rationale for writing this book*.

While this book is making the case for moral obligations to future generations and for a corresponding transformation of politics, its starting point (but also no more than this) remains Hobbesian, as it gives paramount importance to the survival question, that is, to the ultimate question of life and death as far as it depends on the *polis*, that is, the form we give to our association or community, even if *polis* may now turn out to be a *cosmo-polis*. The "idealistic" or "cosmopolitical" conclusion, a concern for humanity,[20] argued in this book should not deceive the reader on how seriously I am taking the lesson of political and anthropological realism; if we are now beyond realism (and idealism), it is not because of a doctrinal choice of the author, but because global challenges are taking us there. I have highlighted in chapter two, § 2, the importance of the "Hobbesian moment" in the belief that, although since Hobbes the anthropology of the individuals and the character of the state have been considerably transformed, individuals and states have not been so civilized or moralized (in Kantian parlance) as to become as sensitive to the question of justice as they were and are to those of survival. A true turning point, if any, in the nature and structure of political association is thus likely to stem from a new stage of confrontation with *life-and-death questions*. The novelty that I am trying to make evident here is that, just because they are under an unprecedented threat, individuals and states have now chances and reasons to become civilized and morally mature enough as to take care of life and death not just for themselves, but for their posterity as well. They do so not because they are now heeding a deeper sense for justice and fraternity, but because they are more closely confronted with the possible self-inflicted death or ruin of themselves and their successors. They may in this context develop higher moral sentiments and come to love each other and to pursue reciprocal fairness, but this is another chapter and is not the initial step or the independent variable. This explains why I reject any generic understanding of global challenges as issues of potential global concern such as AIDS, famine, or tsunami, which can be regarded as global only on grounds of a particular moral or political value judgment, the former related for example to justice or solidarity, the latter to social stability as a premise to a reasonable security policy. As pointed out in the introduction to this book, nuclear weapons and global warming are not chosen among other candidate threats as examples representing the others that are let out. As long as other threats do not reveal in verifiable terms the same moral and political characteristics highlighted here, the *same degree of duress and urgency*, the two global challenges we have been talking about are the only ones existing, period. All the others lack the "Hobbesian" impact on life and death as issues for the modern polity that we have been working out in their case and remain *moderate* challenges, however new their worldwide dimension may be.

Having again justified the restriction to the two items, let us go over to the institutional change they may set in motion. In both cases we modern citizens are used to believe that an ultimate threat can only be checked by a supreme authority possessing not only the key to a benevolent solution (elimination of nuclear weapons or their concentration in one end; substantial emissions cuts and just distribution of them among countries), but the executive force to implement it against everybody as well. In the previous discussion we have looked at the supreme, but illegitimate authority of a realistically unlikely empire. We have hinted at the eventuality that the empire may later on gather legitimacy just because of the protection it can offer against the global challenges. But in the time after 1945 the institution of choice capable to provide safety is generally expected to be legitimate in its essence, and not a posteriori because of its performance, since it is assumed to derive from the free democratic will of states and citizens, as in the design of the UN Charter. To counter the "scourge of war" and particularly the nuclear threat, a voluntarily built-up *world state*[21] is deemed necessary.

Critical remarks on this notion, in order of ascending relevance, regard its redundancy, its legitimacy, and its being based upon the "domestic analogy." Let us first deal with the *redundancy* objection. If the problem is inter-state nuclear anarchy, this can be eliminated by conferring all existing weapons to a central authority, which should be able to detect early enough and to neutralize all attempts by individual states or alliances to rebuild a nuclear capability of their own. This does not need to be done by a state. A sound application of the subsidiarity principle[22] shows that a Supranational Nuclear Agency would be enough, provided it is endowed with an overwhelming political and military power. It would just deal with the nuclear anarchy issue, instead of extending its job all over the world society and its several areas of public affairs; it would certainly exert a tremendous power on any single state, putting essential restraints on its (however suicidal) external sovereignty, but it would lack the oppressing universal presence of a world state and better accommodate the wish of the individual states and societies to maintain their autonomy as far as it does not prejudice the effective elimination of an anarchy leading to omnicide.[23] As to the global warming issue, we are still on a fictional terrain, as in the mostly intergovernmental framework of the discussion around the Kyoto Protocol, a similar *supranational* Agency for Emissions Control (with majority voting) is not supported by anybody.[24] Nor does the intellectual debate on a world state make specific and pointed reference to its role in emissions reduction. But, among the futuribles we cannot do without thinking about in a civilization of rapid transformations, it is possible to imagine a situation in which the concern for global warming and climate disruption become as strong as to raise the request for worldwide control, accompanied by sanctions against non-cooperative or free-riding countries. A transformation of the UN Environmental Program in supranational sense would become possible, perhaps even more than the creation of a Nuclear Agency, as accepting the burden to reshuffle their economy in an environment-friendly way would be less unacceptable to the states than giving up their military autonomy. Also, economic sanctions are likely to be more effective and viable against polluters than nuclear evil-doers, who should be rather hit by diplomatic and military sanctions that cannot be easily decided upon.

Institutions such as the mentioned Agencies, and even more so a world state, would have a threefold problem of *legitimacy*. This has to be seen under three regards:

- goal definition: transnational public opinion as well as statesmen should be in their majority convinced of the urgency to address one and/or the other global challenge as a common enterprise;
- goal achievement: the Agencies need to work effectively and to show visible results, otherwise the objections against their existence would overrun their acceptance. However, even the "output legitimacy"[25] that could be obtained in this way (a case of governance *for* the people) would not eliminate the need for
- representation: by definition the Agencies would have supranational or, in EU parlance, communitarian authority, based on majority voting in the decision-making organs, but the problem remains open whether states or citizens or a mix of them should be represented in those organs. A degree of governance *by* the people is needed to ensure legitimacy.

This last aspect of the legitimacy problem allows for a first objection to radical cosmopolitanism. Be it a world state or a worldwide special, for example, environmental institution, cosmopolitical authors insist on the necessity to put them on a worldwide electoral basis.[26] They miss several important points:

- Free elections do not exist in wide regions of the globe nor will be established in the coming decades, either because elections are not free (for example in China and other authoritarian regimes) or because they cannot be organized in widely illiterate or nomadic societies (some African countries). It would be unfair and unwise to confer the prestige and legitimacy of a worldwide House of Commons upon a parliament in which 1.3 billion Chinese citizens and several other hundreds of millions would be misrepresented or not at all represented.
- Particularly in countries with weak or no democracy the people would be represented in a populist rather than democratic way. Demagogues would draw electoral force from the resentment against better-off peoples and make the international/supranational institution a scapegoat for local failure in governance, as the present phase in EU-politics shows.[27] This can be balanced in countries with democratic traditions, but would perhaps find no containment in countries in which elections have been recently introduced without democratic framework (respect for rule of law and fundamental rights, toleration, break with the dependence from tribal or religious authorities). To require the election of a world parliament means ignoring these crucial political and cultural factors, not to speak of the technical and procedural ones, and making *democracy a mockery*, as though democracy could be exclusively identified with the majority rule, which is just one of its components. Or it means projecting on the entire world an Eurocentric view, as would everybody on the earth approach politics and democracy exactly in the same way as Europeans (or Americans or Canadians or Australians) do.
- An intergovernmental basis for the Agencies, should they ever take shape, is far from satisfactory but for the time being (decades, possibly) preferable for two reasons. First, the inability of certain governments and parliaments to really represent the people would remain open to international checks,[28] a circumstance that would help democratic mobilization in their own countries. This would be not the case if the country were to have sway in the Agencies whatever its regime. Second, global challenges are touchy issues requiring accurate information, unemotional negotiation and a sharp sense for the consequences of policy, as a single wrong decision can lead to disaster. Under the present condition of democracy in the world, intergovernmental negotiations, perhaps stimulated by a worldwide parliamentary assembly with consulting role on global challenges, seem more adequate to bring the parties closer to each other and to prevent hasty steps than a world parliament.

3. TWO VERSIONS OF OBSOLETE STATISM

Before we go on discussing the second and third solution, let us have a detour across the conceptual fundaments of all proposals of world state, supranational agency, and enforceable legal order for all dwellers on the earth. They are all subject to indictment for "domestic analogy," that is, for applying to the solution of the problem of violence between political communities the same theoretical and normative criteria that we are used to when we look at how the modern state has rooted out generalized violence among individuals.[29] *Analogical thinking*, although an indispensable tool, cannot be applied without critical check, which in this case tells us, as Hobbes already knew, that human beings and political communities are too different from each other as to allow for an analogy. In this sense "domestic analogy" is the guillotine under which all talk of a world state in contractarian terms ends up. On the other hand, the question can be raised how far the unprecedented nature of global challenges can debunk that guillotine-like argument, once states and even humankind appear to be subject to the same lethal threats as individuals in the state of nature. In the absence of a superior power capable to enforce peace among the superpowers, in the Cold War situation known as the balance of terror it was indeed the fear of destruction that fulfilled the function of an impersonal and psychological rather than political *tertius*. Will fear of nuclear war and disastrous climate change be strong enough and push in the same direction (a Superleviathan, this time among states) as the fear (cf. chapter two, § 1) experienced in the Thirty Years' War or the English and Scottish civil wars and revolutions of the sixteenth and seventeenth centuries?[30]

The *future of fear* is a complex matter. Perhaps after shocking events such as a regional nuclear exchange and the subsequent wave of sorrow, or Manhattan getting under seawater, it cannot be excluded that it works reasonably in the Hobbesian protecting way, allowing for the creation of a common power. Not to forget however is that fear can also develop in destructive or disrupting forms such as panics, mistrust, or hatred against minorities, as it sometimes happened in America after 9.11 (cf. Kakutani 2001). Regardless of this, four major factors work against its reasonable development.

First, in order to generate the appalling thought of the consequences deriving from a nuclear war or a critical threshold in global warming a fairly high level of education, information, and detachment from basic individual needs (food, job) is required that on a larger scale is given only in the developed countries and in general among elites. Also in these countries less abstract, more tangible and present preoccupations are felt among the public and stimulate decision-makers to take action (joblessness, terrorism, crime). Fear can hardly been talked about as a general emotion; if we want to focus on its political effects we must also reply to the question "whose fear?" Second, a type of fear that is expected to bring human beings closer on a planetary scale may work pretty differently and less powerfully than the fear that made them join together in a limited community, the modern state, for which some identifying elements (borders, cultural affinity, existing communitarian life) pre-existed, while the psychological ties created by economic and cultural globalization are still incomparably weaker. Fear may work as a *tertius*, but so far it did so as force with paralyzing effects on tendencies to wage a nuclear war in a split world of reciprocal terror; it is not said that it can as well work as a unifying creative force leading to the establishment of a common institution. Third, as already mentioned,[31] fear is no longer what it used to be at the creation of Leviathan. On the side of what is feared, the incommensurable destruction that could be brought about by global disasters is in itself likely to set in motion either denial mechanisms or fatalistic acceptance, thus neutralizing all pro-active effects of fear. On the side of those who experience fear, particular in the Western elites, the narcissist concentration on one's own Self and the subsequent lack of interest for the future and collective action, the link between "feeling" fear and "acting" on it in a reasonable and collective way is severed in this case too. Further, among non-Western elites, but also in Western societies, feelings of insecurity generated by social disruption or external threats can breed communitarian tribalism rather than rational fear. Fourth, fear of global disaster and the resulting choice for a protecting Superleviathan are in conflict not just with diverging (neorealist) notions of security or short-termed security interests, but also with the fear of tyranny and the fear of the leveling of diversity. The first one was theorized already by Kant,[32] while the second has come up with more relevance along with the implosion of the "modern" rationalist dream of big systems, the collapse of totalitarianism and the rise of feminism.

The future of fear contains questions of two different levels. On the level of prediction, a Superleviathan or Agencies motivated by fear are conceptually possible, as we have seen, but not likely to come up because of the counterforces. Which trend will prevail is an empirical question that philosophy can pretend to decide only by vainly overreaching. On the normative level, the fear of tyranny and the fear of the leveling of diversity must be acknowledged as rational and protective and kept in a balance with the like-minded fear of nuclear or climate-related disaster, because they all result from a concern for the survival of humankind under humane and acceptable conditions.[33] For this same reason they *can* be kept in a balance, and it should be *possible* to avert the awful situation in which humankind has to face the alternative "either worldwide tyranny or all-out nuclear war/climate out-of-control." That this is conceptually possible does not mean that it is likely or guaranteed, because, as Frederick the Great of Prussia[34] put it when philo-sophically musing on history, every generation wants to make its own stupid mistakes, instead of learning from those of the previous generations; and it is far from proved that the present generations have learnt from the recent past and are ready to forgo their chance of stupid mistakes, which could however be the last ones in history.

This is why, even if theoretically flawed, cosmopolitanism based on the fear mechanisms assumed in the *domestic analogy* is more convincing than any notion of world government appealing to the sense of humanity that is as-sumed to be present in the human beings and only needs to be reactivated by humanistic or religious hortatory. Men (and women) have come closer to self-destruction not because they are wretched, and need to be converted, although they are not naturally good and altruistic either, but rather because they have found no better solution to their security needs in a lethal situation created by a technology they have themselves developed, however not aware of the global disaster it could bring about, be it in war, be it in the exploitation of nature. Structural realism (cf. Donnelly 1992) is a better starting point than an overall pessimistic or optimistic anthropology, because it takes us straight to the point and does not require us to engage in a general philosophy of man straddling the line between good and evil.

A last word on world state proposals and the related domestic analogy: as it happens, their main flaw is their strength. By assuming that before today's problems protection and well-being of all men and women on earth must be provided by a planetary extension of the modern state, this kind of cos-mopolitanism does not take note of the deep changes undergone by politics such as the crisis of sovereignty, the alternative between and the mix of gov-ernment and governance, the new importance of horizontal networks of insti-tutions compared with the pyramidal state. It is an obsolete and statist con-ception of how to get out of the present troubles, which, while lacking the perception of the new trends and chances, resorts to old man Leviathan and must fail, for there is hardly anything similar in sight. But exactly because it is a traditional, well-known, and until recently reliable view, the idea that even in the worldwide dimension *only a state* can set peace, protect life, and stop aggression will last long. Furthermore, this view will always have an edge on the political arrangements of the future, none of which (except a very unlikely planetary empire) will effectively provide and, what is more, give the impression of providing the same degree of security which the territori-ally self-contained state used to provide, and is still now craved for, but plainly impossible to generate in the world of nuclear weapons, global warm-ing, international terrorism, and rapid economic change.

Not so odd is perhaps the announcement, made at the outset of this section, that in the nuclear era the sworn enemy of normative cosmopolitanism, political realism,[35] is affected by the same lack of imagination and sticks to a similar *obsolete statism*. It is not so odd for the reader because this book is permeated by the opinion that pure normativism/idealism/liberalism and "hardshell realism" (see Paul 1997) are relics of a past modernity rather than timeless positions that can apply to the nuclear and global era as well. The best known document of the approach to global challenges is Kenneth Waltz's policy suggestion to let nuclear weapons proliferate because "more may be better," because "nuclear weaponry makes miscalculations difficult" and war less likely, as "new nuclear states will fell the constraints that present nuclear states have experienced" (Waltz in Sagan and Waltz 2003, 44). The background idea is that "deterrence does not depend on rationality. It depends on fear. To create fear, nuclear weapons are the best possible means" (Waltz in Sagan and Waltz 2003, 154).

While I share the counter-arguments presented by Scott Sagan and others, but will not rerun them here, I wish to stress the theoretically significant aspects of this policy controversy. First, Waltz's threat perception is limited to the post–Cold War international system, it does not raise the question of what the unchecked permanence of nuclear weaponry as a normal attribute of sovereign statehood may cause in a far future, or what may follow from a serious attempt of the present generations to neutralize that weaponry. Admittedly, it is just a chance, since, even if we disarm, our posterity will be able to resort to nuclear war technology; but a moral and political turning point like a broad disarmament initiative could have a heavy influence on future choices. Second, it is unjustified to assume that in the future all proliferating states will be able to manage fear in the same rational way as the Western, the Soviet/Russian, and Chinese elites have done so far; the latter having been educated in Marxism-Leninism, an offshoot of European secular mentality. Already in our days the Indian and Pakistani arsenals could fall in the hands of ultra-nationalist or fundamentalist governments, the North Korean bombs are by their very origin just the life insurance of an Oriental satrapy, while the president of the Islamic Republic of Iran, Ahmadinejad, is credited[36] to expect soon the reappearance of the twelfth Imam or "hidden Imam," which according to a Shia prophecy will occur amidst war and destruction as transition to universal peace. Since religious fundamentalism is likely for decades ahead to have a permanent place among the forces influencing international politics, it is hard to see how nuclear states with this kind of regime can be assimilated in their expected behavior to old proliferating countries such as China or Israel. Unlike in realism's basic dogma, in the present world society states are not like states, as their different relationship to their own society or culture (see Evangelista, 1997) can determine a fairly different behavior among them, far away from the patterns we inherited from Western history. A further transfer of old statist patterns from the West to the world under global challenges can perhaps be seen in the implicit persuasion that multipolar deterrence among scores of states will work in the same way as the balance of power system among European states did until 1914; or indeed even better, because for this school of thought deterrence will replace war as a stronger stabilizer of the system. Finally, the realist proponents of "more may be better" are unperceptive of the deep transformation undermining statehood and power in international relations that are for example signaled in Bertrand Badie's formula of the "powerlessness of power" (2004).

Due to the obsolete grounds on which its arguments are based, the idea that "more may be better" seems to be out of touch with present and future realities and deserves the label of "utopian realism" (Betts 2000). No more convincing is the likewise utopian proposal (a policy suggestion rather than an articulated doctrine) to let the market alone solve the problem of global warming. If taken seriously, that is, not as a mere polluting license for fossil fuel consumers, it reveals a *market fundamentalism* which ignores the basic knowledge that markets do not exist without regulation, that the problem is not "market freedom vs. public regulation," but rather how much regulation, and where from (nation-state, binding international agreements, or else), is necessary to let the market orderly work. The limits to regulation are supposed to be set with regard to the balance of the advantages and disadvantages deriving to the society from the various degrees of regulation. But the market, typically a business among actual interest bearers, is not equipped to take care of the interests of market participants who are still far from being born. This is a further reason why in front of global warming market governance cannot be totally entrusted to the market internal mechanisms. Things do not substantially change if the focus of the proposal shifts from market to technology, which is also sometimes credited to provide the solution to global warming. Technology does not come of itself into being, but is generated by public policies in the research sector or by appropriate regulations of the market and corresponding stimuli to private research. An isolated and reified notion of market or technology is unlikely to bring about the solution for the troubles (unregulated growth, private consumption, and pollution of public goods such as the atmosphere) that the not completely lucky encounter between those two factors created.

If because of partly different and partly similar reasons the strategies addressing global challenges that are proposed by cosmopolitanism and realism are not convincing, what else can be done? Is the third way we hinted at above really available?

4. SIGNS OF CHANGE

To answer the strategic question, it is helpful to recapitulate the problem we are facing. It is a collective action problem: how to motivate actors to join under a common rule, if these actors are the states which have so far relied on sovereign self-help in security policy and asserted the priority of each one's unlimited economic growth with regard to nature, technology, and future generations. It is moreover a *collective action problem of new type*: the collective action required to meet global challenges should come not from a bunch of particular actors, but from a universal one, humankind, who against what is common use in politics would take care of the interest of generations of the far future, provided we accept the reasons presented in chapters five and six for merging our and their interest. As noted several times in this book, only ultimate "Hobbesian" threats such as nuclear weapons and global warming can inspire so much concern to persons of good will, and so much fear to everybody as to generate the constraints and the incentives for collective action. By constraints I mean for example the enforceable rules of a strict non-proliferation and disarmament regime, by incentives the expectation of diminishing fear and insecurity and receiving increased cooperation and international aid for those joining a reasonable nuclear weapons or emissions regime.

Such a collective action is clearly not possible with most of the states as they presently are. Can they change? Existing partial exceptions (the EU, Japan, and Russia positive on ratifying the Kyoto Protocol, the Latin-American states sticking to the Treaty of Ttatelolco banning nuclear weapons from the region) show that a process of change is possible and underway, but optimism is unjustified: the process of change can fail to involve a sufficient number of states, it can be paralyzed by a wave of renewed nationalism[37] or tribalism in significant countries, and, what is more, can be accomplished too late, after a critical threshold has been passed (in the inertia effect of global warming or in the brinkmanship of new nations going nuclear).[38] In any case, it would be deceptive if we would now assemble lists of states that are changing and those that are not, as states are complex and ambivalent actors,[39] beyond not being the only actors in world politics. More telling are the *elements and processes of change* and the *signs* of these processes that we can tentatively pinpoint.

First and fundamental element is that, very much unlike in the early modernity, political action comes now into being under conditions of *reflexivity*: we know what the threats in the short and in the long haul are, we are warned that out of our ill-advised present actions nuclear devastation or devastating climate change can result in the future. This is an important cognitive precondition[40] for taking responsibility and for learning processes among individuals and institutions. *Learning* is here understood in the sense defined by Ernst B. Haas, as a situation in which "an organization is induced to question the basic beliefs underlying the selection of ends" (Haas 1990, 36). Unlike technical adaptation to new circumstances, it reexamines beliefs of cause and effect and enables actors to manage interdependence. Under conditions of reflexivity it is thus possible to build up international institutions in a way based on a shared design[41] of what they are for, not only now, but in the future as well. The challenge to Leviathan (the polity) can be taken up with better chances than the obsolescent Leviathans (the modern states) have ever had in the face of problems they were and are not up to.

Institutional design does not mean however to set up a world government or supranational agencies or a worldwide *Comité de salut publique* of Jacobin make. This cannot be ruled out, but it is as well possible that global challenges exert a Hobbesian fear of ultimate danger without necessarily leading to an updated planetary Leviathan. To this purpose the existing patchwork of partial institutions, which provides some governance to world affairs and will probably remain for decades ahead the only practicable way to do so, should be redesigned in a more effective way, with a new political will and animated by a politically cultural awareness of what is at stake. The present system of governance[42] also contains some pieces of government: the UN Security Council, the G8, the World Trade Organization, and a regional semi-polity, the European Union, which are indeed fora for negotiation, communication, and in the best case reciprocal learning rather than authoritative decision-making bodies (with the exception of the more politically integrated EU). Very important for the agenda setting in world politics are formalized epistemic communities such as the International Atomic Energy Agency and the Intergovernmental Panel on Climate Change, which, though not supranational in their definition, do not work in the sense of intergovernmental bargaining; their staff are rather humankind's civil servants.[43] Difficult to assess, but undeniable is the role of public opinion movements and environmental advocacy groups.

More important than single institutions is the whole network of formal and informal (regimes) institutions and its capability to launch and implement policies addressing global challenges, in a word to produce *governance*.[44] Which mix of formal and informal, intergovernmental and supranational institutions is best, how much government is needed for good governance are theoretically non-decidable questions, as they are variables dependent on what is best at any given moment in order to ensure denuclearization and emissions reduction. As a kind of regulative idea in the Kantian sense, the *appeal of old Leviathan* and the domestic analogy mechanisms (cf. p. 199) will still let us think for a while that the world would be nicer and safer if there were a central institution in permanent charge of those policies; but this picture is likely to remain a nostalgic touchstone for the performance of the governance network, not a real alternative to it. Among the several reasons for this, one and not the least is that coercive power plays a less determinant role in world affairs, even if it remains decisive in no-exit situations and always important in carrot-and-stick games. In other cases legal international norms can be usually implemented and "enforced" by non-deadly sanctions and incentives of economic or non-economic (recognition, prestige, and other civilized forms of Hobbesian "glory") nature.[45] This is not the only reason to highlight the high relevance of international law for our concluding argument.

Some legal novelties regarding "mankind" are a sign that the chances for *humankind*, including future generations, to be protected from global threats are now not left to strength of the ethical and philosophical arguments illustrated in chapters five and six, but do also rely on existing, if initial and limited, diplomatic and juridical developments. Already the Outer Space Treaty of 1967 stated in Art. I that the outer space shall be "the province of all mankind." Later on, in the Law of the Sea Convention of 1979 and the Moon Treaty of 1982, the concept of a "common heritage of mankind" was introduced, while the less binding notion of a "common concern of mankind" was used in the Resolution 43/53 of the UN General Assembly of 1988 and in the UN Framework Convention on Climate Change of 1992 (the Rio de Janeiro Convention) with regard to climate change.[46] Humankind has thus become a notion contained in binding international law and referred to indivisible (climate) and divisible (seabed, ocean floor, moon) objects, and this has happened as an answer to problems and chances generated by huge technological advancement. Signs of change also can be seen in jurisprudence, as in the ten years old finding of the International Court of Justice, which denies in general, though not absolutely the legality of the use of nuclear weapons.[47] Only a fraction of this normative activity has real impact on the behavior of states and other actors; for example it has very little effect on emissions cutting, except in the European Union, due to its internal governance. But an ongoing change in legal ideas is a piece of change of the political culture, a powerful element of new governance. Further, the change we have mentioned needs to be seen in a link with the legal phenomenon called globalization of human rights; this can be seen as an effective resurrection and expansion of Kant's idea that human beings as such have some fundamental rights that ought to be recognized by all states and can be enforced by states and/or international institutions ("humanitarian intervention," as in the former Yugoslavia in 1995 and 1999).

If it is true that to the handling of global challenges governance is more relevant than government, and that governance can be tested on the basis of the policies it implements, it is also true that policies stem from *political will*, and we would have then to focus on how a political will addressing the nuclear and the environmental threat could come into life. This cannot be possibly made here, because it would require to get into the political economy and the political sociology of at least ten countries and innumerable transnational networks; and also because it would mean much speculation on things to come. I can only stress that an essential condition in generating a new political will is a *change in the political culture* of elites and voters, which unlike general cultural change makes policy change immediately possible. Debunking the myth of timelessly safe deterrence among few or many countries, binding together the interest of the present and the far future generations in a livable ecosphere, exposing generational nepotism or the denial mechanisms that play down our vulnerability are all intellectual operations that must accompany the birth of a new political will. It is of little help to hypothesize how to let the actual interests of the players coalesce or political forces and leaders come together if the mental environment has not sufficiently changed. Ideas matter in the reflexive policy-making on global issues, even more than they do in traditional foreign policy.[48] This book is intended as a contribution to this change, and this is, if any, its only practical usage.

In what is changing and what has to change the *identity* question is crucial. Put in classical terms and in spite of all suspicion of domestic analogy, do global challenges work like the Hobbesian fear for life and limb? Do they have enough strength as to let us feel all in the same boat, beyond all difference and former enmity, and in need of a new and global protector? In philosophical language, are we going to feel like members of a *non-voluntaristic community*, sharing a fate we cannot escape and therefore bound to act together and to be subject to the same inescapable decisions? Since only a community like this can be strictly called political, is humankind on the way to become a political community, a true global polity?[49] The first answer is: it is possible, global challenges have this potential, whether or not it will happen cannot be safely predicted, there are however many opposite forces and many difficulties, it could as well go awry. The subordinate answer is: even if it happens, it will unlikely be a new institutional Leviathan, but a more complex thing, in which identity will in any case be crucial, because if the dwellers on the earth do not feel like "us," all decisions made on their behalf by the world government or the global governance network to solve global challenges will lack *legitimacy*. This category remains a substantial element in a polity of whatever dimension, and political identity, in the narrow sense mentioned before, is an indispensable factor of legitimacy, particularly in a democratic context.[50] It must be specified that a *political* (not moral or religious) identity as members of the endangered humankind is not expected to eliminate or neutralize our regional or national or transnational (e.g. European) identities, which can rather form concentric circles (cf. Cerutti 2001). Last not least, re-structuring and broadening our multi-layered identity means restructuring our interests as well, as underlined by constructivism and, in another perspective, liberal institutionalism. Countries do not proliferate or start cutting emissions just because they are intimidated or enticed with incentives to do so, but also because their elites see the necessity to rethink their own worldviews and the interplay of particular and universal dimensions and allegiances.

A last recapitulating word on the role of institutions in the face of global challenges that has been so particularly highlighted all throughout this book. Their particular shape (world government, single-issue agencies, networks, and regimes) is obviously relevant to their ability to impact the solutions, but should not be overstated and reified. Even more than their internal design, at the end what matters is rather how far they are led to feel the urgency to ad-dress the two challenges we have analyzed, the only ones that can push them towards acting as effective providers of (however relative) protection for hu-mankind.

5. POLICY RECOMMENDATIONS AND THE LIMITS OF PHILOSOPHY

It is old use to conclude a book on political matter offering some policy rec-ommendations. In the first place I will not evade this obligation, but I will end up questioning it.

The first policy coming in mind about the threat of nuclear weapons is their *abolition*.[51] As far as it is proclaimed as the immediate translation of a moral duty into a policy, it falls within the rejection of any confusion between morality and politics and of the useless naïveté of pure normativism which permeates this book. Radical proposals may satisfy the moral self-esteem of their author and addressees; but they rarely live up to the deeper moral crite-rion that asks scholars of politics to propose solutions which are truly ade-quate to the discouraging intricacy of politics and the intellectual requirement to indicate not just proposals that look good, but also social forces and his-torical tendencies capable of endorsing them. The counterpart of abolition on the side of climate change would be to ask straightforward for a strong re-duction of economic growth and consequently energy consumption around the world, for which no support is available among countries and social groups, while it would also mean fewer chances for the poor to get out of poverty (not the reduction, but the quality of growth seems to be rather the point).

What could abolition mean, if we want to make sense of it? Not an event, on which all nuclear states lay down and dismantle their warheads; not a world in which nuclear systems will be totally absent, not only because after a disarmament agreement cheaters and free riders could secretly rearm themselves, but primarily because disarming states may request a life insurance consisting of, say, a nuclear-armed Security Council capable of deterring everybody from cheating and free-riding. That states renounce self-help including the appropriate arsenal is difficult, though not impossible, because self-help may finally appear to be too costly in the sense of the nuclear security dilemma (see chapter three, § 4); for sure it is not going to be a thing of tomorrow. Should the renouncements ever come about, states may still not want to rely exclusively on reciprocal trust and look for a minimum deterrence force in neutral hands.

That nuclear technology as every one else cannot be disinvented and would always hover over our posterity even after abolition is true, but cannot be used as an argument against universal nuclear disarmament, a formula that I prefer over "abolition." Once production and maintenance of nuclear arms are shut down, only the basic scientific statements would survive, such as the Teller-Ulam configuration for the H-bomb, while the relative technical knowledge would be lost, and it would take time to rebuild it.[52] But in our time universal disarmament will at best go on down to minimum deterrence, and no further; it is premature to discuss the policies implementing complete disarmament as long as the technological (will there be new weapons?) and political circumstances (what will power relationships among states, what will global governance be like at that moment? what developments pushing towards abolition?) are not known. It is also important to note, as Butler Lee, a former commander of the U.S. Strategic Air Command, put it, that the risks of abolition would be different and arguably much lesser in the final stage of the process than they look to be in the present situation of attempted proliferation and obstinate if toothless reaffirmation of national sovereignty (Lee 2000). As all motivating words, "abolition" must simplify things and can turn simplistic; on the other hand, once states had agreed on this perspective, it would in part produce itself conditions easing its implementation (MacKenzie 1999, 433). As for now, abolition does not seem to have chances as a policy, while other policies that have the effect of making the environment more favorable to it do:

- a "no first use" convention
- the de-alerting of nuclear systems, which would run contrary to entrusting security decisions to automatic technologies (cf. chapter three, § 3)[53]
- regardless of its not too numerous failures (in forty years: Israel, India, Pakistan, possibly North Korea and Iran), do not relax the non-proliferation regime and re-legitimate it by finally taking seriously Art. IV, 2 (states in a position to do so shall help non-nuclear, especially developing states in

pursuing the peaceful application of nuclear energy) and above all, as the
Canberra Commission underlined, Art. VI (all parties to the Treaty shall
work for a treaty on general and complete disarmament) of the Non-Proliferation Treaty. The regime cannot live only of monitoring and punishing,
and needs to be nurtured by *political* means, the lack of a general negotiated process towards disarmament having much favored proliferation.

Less exposed to the vagaries of technological change and specific power
relationships are two constitutional policies: governments should establish an
ombudsman for future generations, capable of watching legislation and administrative acts with regard to their assumed impact on the life conditions of
our far posterity. He or she would not be able to block legislation, but, working as an ethical watchdog with some influence on the public opinion, would
be able to question policies suspected to be harmful, thus making their legit-
imization more difficult to achieve. Second, this step would have a greater
impact and a better moral and legal basis if the *right of humankind to survive*
under decent conditions would be recognized as a fundamental right, thus
finding its place in new constitutions or constitutional revisions, and would
bind governments to do their best in that sense and not to act contrary to that
goal. Given the amount in which survival of humankind including future generations has been recently put in danger by human actions, it is hard to understand why the immensely proliferating literature on human rights has
never generated this simple, but symbolically and normatively significant
idea.[54] It should represent the culmination of human rights, now that unlike in
modernity humankind and not only the individuals are at risk and therefore
worth being protected. This constitutional step would also create a favorable
legal precondition for democracy as self-government by the people to claim
the right and the ability of all members of the polity to decide upon their fate.
Doctrines and regimes such as *democracy* can live on only if they prove
themselves able to take up explicitly the new challenges rather than looking
away. This would vindicate sovereignty in its deeper meaning of autonomous
power over the existential issues of the polity, in this case the global polity
whose meaning I have tried to clarify above. For the reasons explained in
chapter four, § 4, the first new challenge to address is the management of the
global commons, protecting them from both a nuclear winter and an unbearable warming, if it is not too late. Under the pressure of the fear for the future
and the concern for posterity the particular polities should become able to
overcome the particularistic interpretation of their interest, joining in a common and disciplined effort, even if they will not probably dissolve in a world
state: admittedly a difficult equation to solve, however not an impossible one.
This effort would be philosophically backed by the new sense of *rationality*

that I have sketched at the end of chapter six and be promoted by the new emotional sense of community based on *solidarity* mentioned in chapter five, § 4. I do however hesitate to describe how more democratic, rational-reasonable, and emotionally cohesive our community could be, once it had decided to live up to the new ultimate challenges. Pictures of a better future risk being consolatory, and modernity is disseminated with failed predictions on what the next state of the world will look like. More conducive to taking action and preventing upcoming evils seems to me to be the inquiry into the mechanism and meaning of those evils, as in Part One of this book, and the discussion of possible counterforces attempted in Part Two.

At this point the reader may feel disappointed: so many pages printed, so many scholars quoted, so many disciplines involved, and this all to generate the very modest proposals that I have just sketched? Where is the project of a better world, where the road map towards salvation that the author's preoccupation with global threats allows hope for?

The author would put on an air of innocence and reply that, as he started the long journey through global challenges, he also heeded secretly some hope, but that the postmodern landscape he was taken through by his exploration induced him then to give it up. He would even note that he formulated his modest policy proposals in the intention not to fully disappoint the reader, rather than seeing in them an essential and consequent practical result of his theory. He does not regret his inability (and unwillingness) to add one more project of perpetual peace to the many that are frustratingly interspersed throughout the long history of European and world wars. The theory developed in this book has strictly speaking no policy outcome, and should not be assessed on the basis of the policies it may be associated with. The reasons for this have been illustrated above in this chapter and need not to be repeated here. The background of this attitude however shall be summarized here as a synthetic conclusion to the whole book, which allows for a weaker and a stronger thesis.

The *weak thesis*: global challenges redefine the modern relationship of politics with ethics and philosophy, but not at all as a way leading back to a unilinear dependence of the former on the latter, since we are prevented from doing so by the lesson of realism as well as by the little encouraging realities of the last hundred years. Politics has no more a full standing on its own, because global challenges overstretch its modern autonomy and make unlikely for it to still be what it used to be, that is, provider of order, peace, and security. Philosophy on the other hand must give up the pretension of designing new worlds, because of its proven intellectual futility, or vice versa of the violent side effects they may unleash or at least legitimate; it should limit itself to make us aware of the problems that lie between past and future, of the perverse consequences of our action, and of the time lag between survival imperatives and the muddy logic of our political action, which can come too late though it is the only one that can match the challenges. Philosophy is no less affected than politics: the pretensions of substantial rationalism are voided as are the monopolistic claims of strategic rationality (cf. chapter three, § 5). The ability to look beyond one's own (generational) nose and to project our interest in survival on posterity is the only substantial, if thinned-down, element of the modern project as universalistic construct that we can retain after modernity has brought about its own disavowal by enhancing a limitless destructiveness. But postmodern catastrophism and consequential stepping-out from world politics is only the reverse image of modernism, incapable of accommodating the basic human interest in survival that is endangered by the outcome of modernity itself.

The *strong thesis* argues that political cosmopolitanism is escapism, which fails to grasp the unavoidable and tragic complexity of politics, particularly in front of global challenges. Unaware of this and other novelties of postmodern politics, which it tends to minimize, is also what we have called realist utopianism. In the future evolution the implementation of a Hobbesian logic of survival leading to the establishment of a Superleviathan cannot be excluded, and can in any case be developed as a mental experiment, but that outcome remains highly unlikely, and not fully protective or without inner risks. Unlike what would happen with nukes, it may not affect climate change in a decisive manner, because of the existing inertia effect in global warming. This is only one more sign that man-made lethal threats will forever overshadow the life and politics of humankind, even if we take timely action. Politics—this is the core of my stronger thesis—can still engineer reasonable institutions of protection, but can no more be credited to be able to shape the good life or to give rise to republican virtues, because both these notions of good old days make hardly any sense in this crude world; while human agency must renounce any pride of itself and any sense of evolutionary achievement. Insofar we are out of modernity and have to look around in order to redefine our place in the world. We may still escape catastrophe, but even if we reconcile with our posterity and with nature, no longer reproducing the dialectics of enlightenment, we will forever live in its shadow, as a perennial possibility of global things going awry.

NOTES

1. I prefer "normativism" because this wording stands clear of the moralistic undertones of "idealism" as well as of the polisemic and confusing features of "liberalism" in its American usage.

2. The personal life expectancy of the members of the dynasty in the foreseeable future is obviously only the human reflex of the political good of stability (*mantenere lo stato*), which was a crucial issue to Machiavelli.

3. Hobbes' refusal of Kantian morality is not weakened by the statement, made in chapter xliii of *Leviathan*, that the article of faith "Jesus is the Christ" is as necessary as obedience to the salvation of man.

4. Kant's thought is paradigmatic for me, too, but I am trying to stick to the real and whole Kant, as he results from his writings, rather than to the oversimplified picture that is sometime used in International Relations literature.

5. Reference is made to Kant 1795, particularly to the First Appendix, *On the Guarantee of Perpetual Peace, passim.*

6. I am hinting at the usage of "morally mature (*moralisiert*)" and "civilized" made by Kant in *Idea for a Universal History with a Cosmopolitan Purpose*, see Kant 1971, 49.

7. To be explicit, under "arrogance" I am thinking of attitudes that can be met in the environmental and even more in the peace movements, while "cynicism" refers to deceptive policies pursued by the forty-third president of the United States in issues of use of force and global warming.

8. Deontologism and consequentialism are indeed more subtly entangled than this clear-cut delimitation may suggest; there are traces of consequentialism even in Kant's foundation of the categorical imperative. But for our present discussion we do not need to further pursue the examination of this entanglement.

9. This resonates with Jonas' "responsibility" as opposed to Ernst Bloch's Marxist "hope."

10. Cf. above on the "scientists' responsibility" at the end of chapter six.

11. This is a variation on the famous saying "all politics is local" by the Bostonian politician Tip O'Neill, the dominating Democratic figure in the U.S. House of Representatives in the seventies and eighties of the twentieth century.

12. In chapter four we have already met a version of this hypothesis in the discussion on how best to cope with global warming, under the captivating neoinstitutionalist label of "institutions for the earth." A more substantial discussion, which cannot be conducted in this book, should obviously include the concept of multilateralism, starting with the formulation introduced by Ruggie 1993.

13. A worldwide monarchy would be "soulless despotism" opposed to the "world republic," the ideal form of federalism that could guarantee perpetual peace, cf. Kant 1795, 113.

14. See Sagan 2000, 37–41.

15. This is probably what misleads anti-Western ideologists to speak of "empire," a construct that has more of political and media rhetoric than intellectual rigor. In more critical terms, the word is used metaphorically by historians when comparing the fate of American power in the world with the crisis of previous empires such as the Roman or the British ones.

16. This was the title of a chapter in Schell's *The Fate of the Earth*, Schell 1982.

17. This story is told in Poundstone 1992, with Matthews' quote on p. 147, and recalled by Scott Sagan in Sagan and Waltz 2003, 55–59.

18. The most recent collection of cosmopolitan perspectives, including representative authors such as Richard Falk and David Held, is Archibugi 2003. I regret not having the space to discuss cosmopolitanism in its various versions, the least abstractly prescriptive versions of which resonate somewhat with aspects of my own approach.

19. See chapter two, § 2, *The Hobbesian Moment.*

20. A broader sense of this notion is to find in Glover 1999, whose subtitle is *A Moral History of the Twentieth Century*, with a chapter on Hiroshima.

21. The last energetic presentation of this claim is to be found in Wendt 2003. Among the many former documents, mostly composed around the end of the Second World War, I shall cite only Einstein's article of 1946 *Towards a World Government* (Einstein 1950, 138–41 and passim).

22. This principle, a major constitutional pillar of the European Union, states that resorting to a higher institutional instance is justified only if the lower ones prove unable to deal with a specific problem.

23. This Agency is what I suggested in my writings of the late eighties–early nineties, as the enhanced cooperation between United States and the reformed Soviet Union/Russia seemed to give a chance in that sense; see Cerutti 1993. While I have meanwhile seen the flaws of any world state proposal, I still consider the Agency more viable than a wholesome world state, should ever things go so far as to make it possible. A similar suggestion was made by Gauthier 1969 in the Appendix on *Hobbes on International Relations.*

24. Only the French President Jacques Chirac, whose mandate expired in 2007, has advanced the proposal to establish a UN Organization for the Environment, which should be capable of coordinated action and have the power to impose sanctions on member states, see Bennhold 2007. However interesting as a sign of a developing debate, this proposal would still lead to an *international* institution, and is probably far from taking shape.

25. Output and input (deriving from democratic representation) legitimacy have been defined by Fritz Scharpf (1999) with regard to the European process, output legitimacy is sometimes credited with a technocratic inclination, which can be the case only if it is seen as the only source of legitimacy. I prefer to speak of Weberian and performance-based legitimacy.

26. See for example Falk and Strauss in Archibugi 2003.

27. I am thinking of the far right and far left opposition to the EU Constitutional Treaty in France in 2005 (cf. Cerutti 2005) and of the ultranationalist coalition still governing Poland in 2007.

28. Minimal liberal and democratic standards would apply for the admission to the Agencies, like those now requested to countries seeking access to the new UN Human Rights Council.

29. Once again I am using a standard view of the domestic analogy argument, originally formulated by the English school of International Relations, and do not get into a differentiated analysis, for which I would like to refer to the standard work by Suganami 1989 and the recent inquiry by Bottici 2004.

30. At the beginning of the nuclear era Reinhold Niebuhr asked if "the degree of mutual danger" will ever be as high as to "overcome the constant inclination of nations to fear each other more than they fear the danger which their enmity has caused for each and both. Were this to happen—he added with skepticism—a really novel factor would have emerged in history." (Niebuhr 1959, 272).

31. For valuable suggestions on the future of fear I am indebted to Michael Brenner and Elena Pulcini.

32. See above in § 2 the question of a " soulless despotism."

33. On the qualification of the survival concern in the context of global challenges cf. chapter five, § 1.

34. Quoted by Koselleck 1979, 46, n. 28.

35. With regard to this sticking to the state the distinction between realism and neorealism does not appear to be relevant; this is the reason why I am speaking in a general sense of realism.

36. According to newspapers articles available in mid 2006 at http://www .expatica.com/source.

37. An example from the recent past: at the beginning (end 2002) of President Lula da Silva's first term, speculations were heard as to the opportunity for Brazil, as the leader of the anti-U.S. attitude in world politics, to launch a nuclear weapons program.

38. Some experts see already proliferation as unavoidable and speak of a post-proliferation world, cf. Rosen 2006.

39. An example: France and the UK stick to their nuclear arsenals, but they do also support Kyoto.

40. Cf. my definition of responsibility, chapter five, § 3.

41. This point is made by Wendt 1999, 376.

42. I am referring to the analytical understanding (as clarified by Pattberg 2006) of the notion of governance as a non-hierarchical, multilevel (spatially and functionally) way to produce global public goods in a new way, not subject to a single organizational principle.

43. This point follows a suggestion made by the late Abram Chayes in a private conversation at Harvard Law School, September 1993.

44. A similar position is upheld by Bellamy and Barry Jones 2000 and Urbinati 2003.

45. Seminal on these developments is Chayes and Handler Chayes 1995. See also Goldstein 2001.

46. See Durner 2001, Baslar 1998, and Genius-Devime 1996. Still interesting is Fasan 1974.

47. See International Court of Justice, Advisory Opinion of 8 July 1996, General List No. 95, available at http://www.icj-cij.org/icjwww/icases/iunan/iunanframe.htm. In § 26 of the Opinion the Court calls the use of nuclear weapons against a particular group genocide under the Genocide Convention of 1948. Notions such as "the human family" and "the heritage of humanity" are found in the Universal Declaration on the Human Genome and Human Rights, adopted by UNESCO's General Conference on November 11, 1997, see http://portal.unesco.org/shs/en.

48. Cf. Goldstein and Keohane (1993), Weber 1997. and Wendt 1999; the latter restricts the role of ideas with regard to the material dimension (biological needs, physical environment, technology).

49. This use goes clearly beyond Modelski's original "weak" definition of global polity (1987) as global political system, concerned with the pursuit of collective action at the global level.

50. On how global challenges impact democratic legitimacy, and why democracy is a better place to restructure identity and interests cf. chapter four, § 4. The link between political identity and legitimacy/legitmation is discussed in my introductory chapter in Cerutti-Lucarelli 2008.

51. Jonathan Schell (1982, 1984) was the leading representative of this doctrine long ago and still is, as can be seen in his insistence on arms control being a folly (2000). Cf. also Calogero 1997.

52. I am taking this argument from MacKenzie 1999, an important article on this topic.

53. For different advice see Deutch 2005.

54. The EU Constitutional Treaty stumbled over quite different matters, but even before the French and Dutch referenda of 2005 its unimaginative and bureaucratic nature was clear, not least because of the lack in it of any attempt to take on the normative problems related to the two global challenges. On the present debate on human rights cf. Kuper 2005.

Works Cited

For the sake of historical perspective sometimes works are quoted according to the year of composition or first edition (e.g. Kant 1795). In this case, the year of the edition I have used is indicated at the end of the bibliographical entry.

Airaksinen, Timo, and Martin Berman, eds. 1989. *Hobbes: War Among Nations.* Aldershot: Avebury.

Aldy, Joseph, Scott Barrett, and Robert Stavins. 2003. *Thirteen Plus One: A Comparison of Global Climate Policy Architectures.* Faculty Research Working Papers Series 03-12, Kennedy School of Government, Harvard Univ. http://ksgnotes1 .harvard.edu/research/wpaper.nsf/rwp/.

Alperovitz, Gar. 1965. *Atomic Diplomacy: Hiroshima and Potsdam.* New York: Simon and Schuster.

Anders, Günther. 1956. *Die Antiquiertheit des Menschen*, vol. 1. Fifth edition. Munich: Beck 1980.

Apel, Karl-Otto. 1988. Verantwortung heute: nur noch Prinzip der Bewahrung und Selbstbeschränkung oder immer noch der Befreiung und Verwirklichung von Humanität? In *Diskurs und Verantwortung.* Frankfurt am Main: Suhrkamp.

Archibugi, Daniele, ed. 2003. *Debating Cosmopolitics.* London: Verso.

Badie, Bertrand. 2004. *L'impuissance de la puissance.* Paris: Fayard.

Bahr, Egon, and Dieter S. Lutz, eds. 1986. *Zu den Ausgangsüberlegungen, Grundlagen und Strukturmerkmalen gemeinsamer Sicherheit.* Baden-Baden: Nomos Verlagsgesellschaft.

Baier, Annette. 1984. For the Sake of Future Generations. In Regan, Tom, ed., *Earthbound: New Introductory Essays in Environmental Ethics.* New York: Random House: 214–46.

Baldry, H.C. 1965. *The Unity of Mankind in Greek Thought.* Cambridge: Cambridge Univ. Press.

Barrett, Scott. 2003. *Environment and Statecraft.* New York: Oxford Univ. Press.

Baslar, Kernel. 1998. *The Concept of the Common Heritage of Mankind in International Law.* The Hague: Nijhoff.

Bauman, Zygmunt. 1999. *In Search of Politics.* Cambridge: Polity Press.

BBC (British Broadcasting Corporation). 2006. *Global Dimming.* http://www.bbc.co .uk/sn/tvradio/programmes/horizon/dimming_prog_summary.shtml.

Beck, Ulrich. 1991. *Politik in der Risikogesellschaft.* Frankfurt am Main: Suhrkamp.

———. 1992. *Risk Society: Towards a New Modernity.* London: Sage.

Bellamy, Richard and R.J. Barry Jones. 2000. Globalization and Democracy—an afterword. In Holden, Barry, ed. *Global Democracy: Key Debates.* London: Routledge: 202–16.

Bennhold, Katrin. 2007. Chirac tells U.S. to join climate protocol or face taxes. *International Herald Tribune,* January 31.

Berger, Peter. 1977. *Facing Up to Modernity: Excursions in Society, Politics, and Religion.* New York: Basic Books.

Berger, Peter, and Thomas Luckmann. 1967. *The Social Construction of Reality.* New York: Doubleday.

Betts, Richard K. 2000. Universal Deterrence or Conceptual Collapse? In Utgoff, Victor A., ed. 2000. *The Coming Crisis: Nuclear Proliferation, U.S. Interests, and World Order.* Cambridge, Mass.: The MIT Press: 51–85.

Blumenberg, Hans. 1983. *The Legitimacy of the Modern Age.* Cambridge, Mass.: The MIT Press.

———. 1997. *Shipwreck with Spectator: Paradigm of a Metaphor for Existence.* Cambridge, Mass.: The MIT Press.

Bobbio, Norberto. 1979. *Il problema della guerra e le vie della pace.* Bologna: Il Mulino.

———. 1981. Stato. In *Enciclopedia Einaudi,* vol. 13. Turin: Einaudi: 453–512.

Bödeker, H.-E. 1980. Menschheit, Menschengeschlecht. In Ritter, Joachim, ed. *Historisches Wörterbuch der Philosophie,* vol. 5. Basel: Schwabe: 1127–37.

———. 1982. Menschheit, Humanität, Humanismus. In Brunner, Otto, Werner Conze, and Reinhart Koselleck, eds. *Geschichtliche Grundbegriffe,* vol. 3. Stuttgart: Klett-Cotta: 1063–1128.

Bottici, Chiara. 2004. *Uomini e Stati.* Pisa: ETS.

Brehmer, Berndt. 1989. *The Psychology of Risk.* In Singleton, W.T., and Jan Hovden, eds. *Risk and Decisions.* New York: Wiley.

Brodie, Bernard, ed. 1946. *The Absolute Weapon: Atomic Power and World Order.* New York: Brace and Co.

———. 1973. *War and Politics.* New York: Macmillan.

Bull, Hedley. 1977. *The Anarchical Society.* New York: Columbia Univ. Press 1995.

Bundy, McGeorge. 1988. *Danger and Survival.* New York: Random House.

Burgman, Mark. 2005. *Risks and Decisions for Conservation and Environmental Management.* Cambridge: Cambridge Univ. Press.

Burroughs, William J. 2001. *Climate Change: A Multidisciplinary Approach.* Cambridge: Cambridge Univ. Press.

Buzan, Barry. 1991. *People, States and Fear*. Second edition. New York: Harvester Wheatsheaf.

Calabresi, Guido, and Philip Bobbitt. 1978. *Tragic Choices*. New York: Norton.

Calogero, Francesco. 1997. A Nuclear Weapon-free World: Is it Desirable? Is it Possible? Is it Probable? In Schroeer, Dietrich, and Alessandro Pascolini, eds. *The Weapons Legacy of the Cold War*. London: Ashgate.

The Canberra Commission. 1997. http://www.dfat.gov.au/cc/cc_report_exec.html.

Care, Norman. 1987. *On Sharing Fate*. Philadelphia: Temple Univ. Press.

Cerutti, Furio. 1993. Ethics and Politics in the Nuclear Age. *Praxis International* 12, no. 4 (January): 387–404.

———. 2001. Towards a Political Identity of the Europeans. In Cerutti, Furio, and Enno Rudolph, eds. *A Soul for Europe*, vol. 1, *A Reader*. Leuven: Peeters: 1–31.

———. 2005. Europe's Deep Crisis. *European Review* 13 no. 4: 525–40.

Cerutti, Furio, and Sonia Lucarelli, eds. 2008 (forthcoming). *The Search for a European Identity*. London: Routledge.

Chaliand, Gérard. 2006. Le terrorisme ne peut pas provoquer de grands bouleversements. *Libération*, March 4 and 5.

Chayes, Abram. 1974. *The Cuban Missile Crisis*. London: Oxford Univ. Press.

Chayes, Abram, and Antonia Handler Chayes. 1995. *The New Sovereignty: Compliance with International Regulatory Agreements*. Cambridge, Mass.: Harvard Univ. Press.

The Chicago Council on Global Affairs. 2007. http://thechicagocouncil.org/media_press_room_detail/.

Choron, Jacques. 1963. *Death and Western Thought*. New York: Collier.

Claude, Inis. 1962. *Power and International Relations*. New York: Random House.

Clausewitz, Carl von. 1832. *On War*. Princeton: Princeton University Press 1976.

Cohen, Joel. 2006. *As a Species, We're Learning to Be a Mature Adult*. http://www.earthsky.org/article/joel-cohen-interview.

Conze, Werner. 1984. Sicherheit, Schutz. In Brunner, Otto, Werner Conze, and Reinhart Koselleck, eds. *Geschichtliche Grundbegriffe*, vol. 5. Stuttgart: Klett-Cotta: 831–62.

Cooper, Richard N. 2006. The Kyoto Protocol: A Flawed Concept. http://www.economics.harvard.edu/faculty/cooper/papers/Kyotoprotocol.pdf.

———. 2006. *Alternatives to Kyoto: the Case for a Carbon Tax*. http://www.economics.harvard.edu/faculty/cooper/papers/Kyoto_ct.pdf.

Crichton, Michael. 2004. *State of Fear*. London: HarperCollins.

Crutzen, Paul J. 2006. Albedo Enhancement by Stratospheric Sulfur Injections: A Contribution to Resolve a Policy Dilemma? *Climatic Change*, http://www.springerlink.com/content/t1vn75m458373h63/.

Dahl, Robert. 1963. *Modern Political Analysis*. Englewood Cliffs, N.J.: Prentice Hall.

———. 1985. *Controlling Nuclear Weapons: Democracy versus Guardianship*. Syracuse, N.Y.: Syracuse Univ. Press.

Davis, Bernard, ed. 1991. *The Genetic Revolution*. Baltimore: The Johns Hopkins Univ. Press.

Delumeau, Jean. 1979. *La peur en Occident*. Paris: Fayard.

———. 1989. *Rassurer et protéger*. Paris: Fayard.

Deutch, John. 2005. A Nuclear Posture for Today. *Foreign Affairs* 84 no. 1 (January/February): 49–60.

Dobson, Andrew. 1998. *Justice and the Environment.* Oxford: Oxford Univ. Press.

———. 2006. Thick Cosmopolitanism. *Political Studies* 54 165–84.

Donnelly, Jack. 1992. Twentieth Century Realism. In Nardin, Terry, and David R. Mapel, eds. *Traditions of International Ethics.* Cambridge: Cambridge Univ. Press: 85–111.

Dotto, Lydia. 1986. *Planet Earth in Jeopardy: Environmental Consequences of Nuclear War.* Chichester: Wiley.

Douglas, Mary, and Aaron Wildavsky. 1982. *Risk and Culture.* Berkeley: Univ. of California Press.

Downing, Thomas E., ed. 1996. *Climate Change and World Food Security.* Berlin: Springer.

Downing, Thomas E. et al., eds. 2002. *Climate, Change and Risk.* London: Routledge.

Durner, Wolfgang. 2001. *Common Goods: Statusprinzipien von Umweltgütern im Völkerrecht.* Baden-Baden: Nomos.

Dyson, Freeman. 2003. What a World! *The New York Review of Books* 20, no. 8, May 1.

Ebeling, Hans, ed. 1976. *Subjektivität und Selbsterhaltung.* Frankfurt am Main: Suhrkamp.

Ehrlich, Paul R. et al. 1984. *The Cold and the Dark: The World after Nuclear War.* New York: Norton.

Einstein, Albert. 1950. *Out of My Later Years.* New York: Philosophical Library.

Eisenstadt, Shmuel. 1968. Social Institution. In Sills, David L., ed. *International Encyclopedia of the Social Sciences.* New York: Macmillan, 14: 409–29.

Elias, Norbert. 1936. *The Civilizing Process.* Malden, Mass.: Blackwell.

Elster, Jon. 1979. Risk, Uncertainty and Nuclear Power. In *Social Science Information*, 18: 371–400.

Evangelista, Matthew. 1997. Domestic Structure and International Change. In Doyle, Michael W., and G. John Ikenberry, eds. *New Thinking in International Relations Theory.* Boulder, Colo.: Westview: 202–26.

Falk, Richard, and Andrew Strauss. 2003. The Deeper Challenges of Global Terrorism: A Democratizing Response. In Archibugi 2003: 203–31.

Fasan, Ernst. 1974. The Meaning of the Term "Mankind" in Space Legal Language, *Journal of Space Law*: 125–31.

Fernandez-Armesto, Felipe. 2004. *So You Think You're Human?* Oxford: Oxford Univ. Press.

Flämig, Christian. 1985. *Die genetische Manipulation des Menschen.* Baden-Baden: Nomos.

Freedman, Lawrence. 2003. *The Evolution of Nuclear Strategy.* Basingstoke: Palgrave Macmillan.

Freud, Sigmund. 1932. Why War? In *The Standard Edition of the Complete Psychological Works*, vol. xxii, London: Hogarth 1964, 197–218.

Fursenko, Aleksandr, and Timothy Naftali. 2006. *Khrushchev's Cold War.* New York: Norton.

Fussell, Paul. 1988. *Thank God for the Atom Bomb.* New York: Ballantine.

Gambetta, Diego, ed. 1988. *Trust*. New York: Blackwell.

Garner, Richard. 2007. All Pupils to Be Given Lessons in Climate Change. *The Independent,* 2 February.

Gauthier, David. 1969. *The Logic of Leviathan*. Oxford: Clarendon.

Gehlen, Arnold. 1980. *Man in the Age of Technology*. New York: Columbia Univ. Press.

Genius-Devime, Barbara. 1996. *Bedeutung und Grenzen des Erbes der Menschheit im völkerrechtlichen Kulturgüterschutz*. Baden-Baden: Nomos.

Gilson, Bernard. 1984. *The Conceptual System of Sovereign Equality*. Leuven: Peters.

Glover, Jonathan. 1999. *Humanity*. New Haven: Yale Univ. Press.

Goklany, Indur M. 2007. *The Improving State of the World*. Washington: Cato Institute.

Goldstein, Judith and Robert Keohane, eds. 1993. *Ideas and Foreign Policy*. Ithaca: Cornell Univ. Press.

Goldstein, Judith et al, eds. 2001. *Legalization and World Politics*. Cambridge, Mass: MIT Press.

Goodin, Robert E. 1985. *Protecting the Vulnerable*. Chicago: The Univ. of Chicago Press.

Gosseries, Axel. 2003. Historical Emissions and Free Riding. In Meyer, L., ed. *Justice in Time: Responding to Historical Injustice*. Baden-Baden: Nomos: 355–82.

Haas, Ernst B. 1990. *When Knowledge is Power*. Berkeley: Univ. of California Press.

Haas, Peter. 1997. Epistemic Communities and the Dynamics of International Environmental Cooperation. In Rittberger, Volker. *Regime Theory and International Relations*. Oxford: Clarendon.

Haas, Peter et al., eds. 1993. *Institutions for the Earth*. Cambridge, Mass.: The MIT Press.

Habermas, Jürgen. 1984–87. *Theory of Communicative Action*. Boston: Beacon Press.

———. 2003. *The Future of Human Nature*. Cambridge: Polity Press.

Hammitt, James K. 1990. *Probability is All We Have*. New York: Garland.

Hanfling, Oswald. 1987. *The Quest for Meaning*. Oxford: Open Univ. and Blackwell.

Hardin, Garrett, and John Baden, eds. 1977. *Managing the Commons*. San Francisco: Freeman.

Hashmi, Sohail H., and Steven P. Lee, eds. 2004. *Ethics and Weapons of Mass Destruction*. Cambridge: Cambridge Univ. Press.

Hegel, Georg W.F. 1821. *Vorlesungen über die Philosophie der Weltgeschichte*, vol. iv. *Die germanische Welt*. Hamburg: Felix Meiner 1968.

Heidegger, Martin. 1977. *The Question Concerning Technology*. New York: Continuum.

———. 1978. *Being and Time*. Oxford: Blackwell.

Henkin, Louis et al., eds. 1980. *International Law*. American Casebook Series. St. Paul, Minn.: West Publishing.

Henrich, Dieter. 1990. *Ethik zum nuklearen Frieden*. Frankfurt am Main: Suhrkamp.

Herken, Gregg. 1988. *The Winning Weapon: The Atomic Bomb in the Cold War, 1945–1950*. Princeton, N.J.: Princeton University Press.

Herz, John. 1959. *International Politics in the Atomic Age.* New York: Columbia Univ. Press.

Hill, Christopher. 2003. *The Changing Politics of Foreign Policy.* Basingstoke: Palgrave.

Hirsch, Günter, and Wolfram Eberbach. 1987. *Auf dem Weg zum künstlichen Leben.* Basel: Birkhäuser.

Hirschman, Albert O. 1970. *Exit, Voice, and Loyalty.* Cambridge, Mass.: Harvard University Press.

Hobbes, Thomas. 1642. *De Cive.* In *Opera, quae latine scripsit omnia.* London: Boiin 1839, vol. II.

——. 1651. *Leviathan.* Harmondsworth, UK: Pelican 1968.

Holden, Barry. 2002. *Democracy and Global Warming.* London: Continuum.

Homer-Dixon, Thomas. 1999. *Environment, Scarcity and Violence.* Princeton, N.J.: Princeton Univ. Press.

Horkheimer, Max. 1978. *Dawn & Decline: Notes 1926–1931 and 1950–1969.* New York: Seabury Press.

——. 1982. The Authoritarian State. In Arato, Andrew, and Eike Gebhardt. *The Essential Frankfurt School Reader.* New York: Continuum: 95–117.

——. 1995. *Critical Theory.* New York: Continuum.

Horkheimer, Max, and Theodor W. Adorno. 1947. *Dialecticts of Enlightenment.* New York: Continuum 1993.

Howard, Michael. 1978. *War and the Liberal Conscience.* Oxford: Oxford Univ. Press.

——. 1984. War in the Making and Unmaking of Europe. In *The Causes of War.* Cambridge, Mass.: Harvard Univ. Press, 151–68.

Hume, David. 1739. *A Treatise of Human Nature.* Penguin: London 1969.

——. 1777. *An Enquiry Concerning the Principles of Moral.* LaSalle: Open Court 1966.

Institute of Medicine. (National Academy of Sciences). 1986. *The Medical Implications of Nuclear War.* Washington: National Academy Press.

IPCC (Intergovernmental Panel on Climate Change). 2001. *Climate Change 2001*: *Synthesis Report. Summary for Policymakers.* http://www.grida.no/climate/ipcc_tar/vol4/english/005.htm.

——. 2007. *Climate Change 2007: The Physical Science Basis. Summary for Policymakers.* http://www.ipcc.ch/SPM2feb07.pdf.

Jamieson, Dale. 1992. Ethics, Public Policy, and Global Warming. *Science, Technology, & Human Values* 17: 139–53.

Jasanoff, Sheila. 1993. Bridging the Two Cultures of Risk Analysis. *Risk Analysis* 13 no. 2: 123–29.

Jervis, Robert. 1976. *Perception and Misperception in International Politics.* Princeton: Princeton University Press.

Johnson, James T. 1980. *Just War Tradition and the Restraint of War.* Princeton, N.J.: Princeton Univ. Press.

——. 1984. *Can Modern War Be Just?* New Haven: Yale Univ. Press.

Jonas, Hans, 1984. *The Imperative of Responsibility.* Chicago: Univ. of Chicago Press.

Kakutani, Michiko. 2001. Fear, the New Virus of a Connected Area. *New York Times,* October 20.

Kant, Immanuel. 1794. The End of All Things. In *Religion and Rational Theology.* Cambridge: Cambridge Univ. Press 1996: 217–32.

———. 1795. Perpetual Peace: A Philosophical Sketch. In Kant 1971, 93–132.

———. 1797. *The Metaphysics of Morals.* Cambridge: Cambridge Univ. Press 1991.

———. 1971. *Kant's Political Writings.* Cambridge: Cambridge Univ. Press.

———. 1988. *Selections.* Englewood Cliffs, N.J.: Prentice Hall.

Kaufmann, Franz-Xaver. 1973. *Sicherheit als soziologisches und sozialpolitisches problem. Untersuchungen zu einer. Wertidee hochdifferenzierter Gesellschaften.* Stuttgart: Enke.

Kavka, Gregory. 1978. The Futurity Problem. In Sikora, Richard, and Brian Barry. *Obligations to Future Generations.* Philadelphia: Temple Univ. Press: 186–203.

Kaysen, Carl. 1991. Is War Obsolete? In Lynn-Jones, Sean M., ed. *The Cold War and After.* Cambridge, Mass.: The MIT Press, 81–103.

Keohane, Robert. 1989. *International Institutions and State Power.* Boulder: Westview.

———. 1992. *Institutionalist Theory and the Realist Challenge after the Cold War.* Working Paper 7, Center for International Affairs, Harvard Univ.

———. 1993. *Hobbes' Dilemma and Institutional Change in World Politics: Sovereignty in International Society.* Working Paper 3, Center for International Affairs, Harvard Univ.

Keohane, Robert, Peter Haas, and Marc Levy. 1993. The Effectiveness of International Environmental Institutions. In Haas et al., eds. *Institutions for the Earth.* Cambridge, Mass.: The MIT Press: 1–24.

Kissinger, Henry A. 2004. Now, Back to Defining a New World Order. *International Herald Tribune,* November 6–7.

Koselleck, Reinhart. 1979. *Vergangene Zukunft.* Frankfurt am Main: Suhrkamp.

Kuper, Andrew, ed. 2005. *Global Responsibilities.* New York: Routledge.

Laïdi, Zaki. 2001. *Un monde privé de sens.* Paris: Hachette.

Lamb, Hubert H. 1995. *Climate, History and Modern World.* London: Routledge.

Larmore, Charles. 1987. *Patterns of Moral Complexity.* Cambridge: Cambridge Univ. Press.

———. 1999. The Idea of a Life Plan. In Paul, E.F. et al., eds. *Human Flourishing.* Cambridge: Cambridge Univ. Press.

Lash, Scott. 1999. *Another Modernity, a Different Rationality.* Oxford: Blackwell.

Laslett, Peter, and James Fishkin, eds. 1992. *Justice between Age Groups and Generations.* New Haven: Yale Univ. Press.

Latour, Bruno. 1993. *We Have Never Been Modern.* New York: Harvester Wheatsheaf.

Lawson, Nigel. 2006. *The Economics and Politics of Climate Change.* www.cps.org .uk/cpsfile.asp?id=641.

Lebow, Richard N., and Janice Gross Stein. 1994. *We All Lost the Cold War*. Princeton: Princeton Univ. Press.

Lee, Butler. 2000. Zero Tolerance. *The Bulletin of the Atomic Scientists*, January–February.

Leiss, William, and Christina Chociolko. 1994. *Risk and Responsibility*. Montreal: McGill-Queen's Univ. Press.

Lelas, Srđan. 2000. *Science and Modernity*. Dordrecht: Kluwer.

Lenk, Hans. 1982. *Zur Sozialphilosophie der Technik*. Frankfurt am Main: Suhrkamp.

Leroux, Marcel. 2002. L'évolution réelle de la dynamique du temps. *Annales de Géographie* 624, 115–37.

Levine, Howard, Daniel Jacobs, and Lowell Rubin, eds. 1988. *Psychoanalysis and the Nuclear Threat*. Hillsdale, N.J.: The Analytic Press.

Levy, Marc, Robert Keohane, and Peter Haas. 1993. Improving the Effectiveness of International Environmental Institutions. In Haas, Peter et al., eds. *Institutions for the Earth*. Cambridge, Mass.: The MIT Press.

Lieber, Robert. 1988. *No Common Power*. Glenview, Ill.: Scott, Foresman & Co.

Lifton, Robert Jay. 1968. *Death in Life: Survivors of Hiroshima*. New York: Basic Books.

Lifton, Robert Jay, and Richard Falk. 1982. *Indefensible Weapons*. New York: Basic Books.

Locke, John. 1690. An Essay Concerning the True Original Extent and End of Civil Government (Second Treatise of Government). In *Two Treatises of Government*. London: Dent 1982.

Lomborg, Bjørn, ed. 2007. *Global Crises, Global Solutions*. Cambridge: Cambridge Univ. Press.

Löwith, Karl. 1964. *Meaning in History*. Chicago: Univ. of Chicago Press.

Lucarelli, Sonia. 1994. La teoria dei giochi ed il suo contributo all'analisi delle crisi internazionali. Studio di un caso: la Alert Crisis dell'Ottobre 1973. *Quaderni Forum*, VI, 4.

Luhmann, Niklas. 1979. *Trust and Power*. New York: Wiley.

———. 1988. Familiarity, Confidence, Trust: Problems and Alternatives. In Gambetta, Diego, ed. *Trust*. New York: Blackwell: 94–108.

Lukes, Steven. 1991. *Moral Conflict and Politics*. Oxford: Clarendon.

Lyotard, Jean François. 1984. *The Postmodern Condition: A Report on Knowledge*. Minneapolis: University of Minnesota Press.

MacIsaac, David. 1986. Voices from the Central Blue: The Air Power Theorists. In Paret, Peter, ed. *Makers of Modern Strategy*. Princeton: Princeton Univ. Press, 624–47.

MacKenzie, Donald. 1999. Theories of Technology and the Abolition of Nuclear Weapons, in Olivier Coutard et al., eds. *The Governance of Large Technical Systems*. London: Routledge: 173–98.

Manning, Martin, and Michel Petit. 2003. A Concept Paper for the AR4 Cross Cutting Theme: Uncertainties and Risk. http://www.ipcc.ch/activity/cc/t1.pdf.

Marx, Karl, and Friedrich Engels. 1845–46. *The German Ideology*. Amherst, N.Y.: Prometheus Books 1998.

———. 1953. *Ausgewählte Briefe*. Berlin: Dietz.

Mead, Margaret. 1964. *Continuities in Cultural Evolution*. New Haven: Yale Univ. Press.

Mearsheimer, John J. 1983. *Conventional Deterrence*. Ithaca: Cornell Univ. Press.

———. 1990. Back to the Future: Instability in Europe after the Cold War. *International Security* 15 no. 1: 5–56.

Merchant, Carolyne. 2003. *Reinventing Eden: The Fate of Nature in Western Civilization*. New York: Routledge.

Miller, Clark A., and Paul N. Edwards, eds. 2001. *Changing the Atmosphere*. Cambridge, Mass.: The MIT Press.

Modelski, George. 1987. *Long Cycles in World Politics*. Seattle: Univ. of Washington Press.

Montesquieu, Charles-Louis de Secondat, baron de. 1748. *The Spirit of the Laws*. Cambridge: Cambridge Univ. Press 1989.

Moore, G.E. 1903. *Principia Ethica*. Cambridge: Cambridge Univ. Press 1993.

Mueller, John. 1991. The Essential Irrelevance of Nuclear Weapons. In Lynn-Jones, Sean M., ed. *The Cold War and After*. Cambridge, Mass.: The MIT Press, 45–80.

Nagel, Thomas. 1987. The Absurdity of Life. In Hanfling, Oswald, ed. *Life and Meaning*. Oxford: Blackwell.

Nardin, Terry. 1983. *Law, Morality, and the Relations of States*. Princeton: Princeton Univ. Press.

Nedelmann, Carl, ed. 1985. *Zur Psychoanalyse der nuklearen Drohung*. Göttingen: Vandenhoeck&Ruprecht.

Neumann, Franz L. 1957. Anxiety and Politics. In *The Democratic and the Authoritarian State*. New York: Free Press.

Niebuhr, Reinhold. 1959. *The Structure of Nations and Empires*. New York: Charles Scribner's Sons.

Nolte, Ernst. 1987. *Der europäische Bürgerkrieg 1917–1946: Nationalsozialismus und Bolschewismus*. Munich: Propiläen.

Nordhaus, William, and Joseph Boyer. 2000. *Warming the World: Economics Models of Global Warming*. Cambridge, Mass.: The MIT Press.

Nozick, Robert. 1989. *The Examined Life*. New York: Simon&Schuster.

Oftedal, Per. 1986. Genetic Consequences of Nuclear War. In Institute of Medicine 1986: 337–48.

Oldfield, Frank. 2005. *Environmental Change*. Cambridge: Cambridge Univ. Press.

Parfit, Derek. 1984. *Reasons and Persons*. Oxford: Clarendon.

Parson, Edward. 1993. Protecting the Ozone Layer. In Haas, Peter et al., eds., *Institutions for the Earth*. Cambridge, Mass.: The MIT Press: 27–74.

———. 2003. *Protecting the Ozone Layer: Science and Strategy*. Oxford: Oxford Univ. Press.

Passmore, John. 1974. *Man's Responsibility for Nature*. London: Duckworth.

Paterson, Matthew. 1996. *Global Warming and Global Politics*. London: Routledge.

———. 2000. *Understanding Global Environmental Politics*. London: Routledge.

Pattberg, Philipp. 2006. Global Governance: Reconstructing a Contested Social Science Concept. *GARNET Working Paper 04/06*. http://www.garnet-eu.org/fileadmin/documents/working_papers/0406.pdf.

Paul, T.V. 1997. *Power versus Prudence: Why Nations Forgo Nuclear Weapons.* In Doyle, Michael W., and G. John Ikenberry, eds. *New Thinking in International Relations Theory.* Boulder, Colo.: Westview.

Perez-Ramos, Antonio. 1996. Bacon's legacy. In Peltonen, Markku. *The Cambridge Companion to Bacon.* Cambridge: Cambridge Univ. Press.

Pew Research Center Survey on Global Warming. 2007. http://people-press.org/reports/.

Philippidis, Leonidas. 1929. *Die "Goldene Regel" religionsgeschichtlich untersucht.* Diss. Phil., Univ. of Leipzig.

Pocock, John, G.A. 1975. *The Machiavellian Moment: Florentine Political Thought and the Atlantic Republican Tradition.* Princeton, NJ: Princeton University Press.

Portinaro, Pier Paolo. 1986. *Il Terzo.* Milano: Angeli.

Poundstone, William. 1992. *Prisoner's Dilemma.* New York: Doubleday.

Powaski, Ronald E. 1987. *March to Armageddon.* New York: Oxford Univ. Press.

Ragionieri, Rodolfo. 1989. *Sicurezza comune.* Fiesole: ECP.

Raup, David M. 1984. Death of Species. In Nitecki, Matthew H. ed. *Extinctions.* Chicago: Univ. of Chicago Press, 3–19.

Rawls, John. 1971. *A Theory of Justice.* Cambridge, Mass.: Harvard Univ. Press.

———. 1999. *The Law of Peoples.* Cambridge, Mass.: Harvard Univ. Press.

Rees, Martin. 2003. *Our Final Hour.* New York: Basic Books.

Rengger, Nicholas. 2005. Tragedy or Scepticism? Defending the Anti-Pelagian Mind in World Politics. *International Relations* 19 no. 3, 321–28.

Ricoeur, Paul. 1990. The Golden Rule. *New Testament Studies* 36: 392–97.

Rosen, Stephen P. 2006. After Proliferation: What to Do If More States Go Nuclear. *Foreign Affairs* 2006, September–October.

Rousseau, Jean Jacques. 1755–56. The State of War. In Hoffmann, Stanley, and David P. Fidler, eds. *Rousseau on International Relations.* Oxford: Clarendon 1991, 33–47.

———. 1762. Social Contract. In *The Collected Writings of Rousseau*, vol. 4, Hanover: Univ. Press of New England 1994.

Rowan-Robinson, Michael. 1985. *Fire and Ice: The Nuclear Winter.* Essex: Longman.

Rueter, Theodore, and Thomas Kalil. 1991. Nuclear Strategy and Nuclear Winter. *World Politics* 43: 587–607.

Ruggie, John G. 1993. *Multilateralism Matters.* New York: Columbia Univ. Press.

Sade, Donatien A.F., marquis de. 2004. *Philosophy in the Boudoir; Minski the Cruel.* New York: Creation Books.

Sagan, Carl, and Richard Turco. 1990. *A Path Where No Man Thought.* New York: Random House.

Sagan, Scott D. 2000. Rethinking the Causes of Nuclear Proliferation: Three Models in Search of a Bomb. In Utgoff, Victor A., ed. *The Coming Crisis: Nuclear Proliferation, U.S. Interests, and World Order.* Cambridge, Mass.: The MIT Press 2000, 17–50.

Sagan, Scott D., and Kenneth Waltz. 2003. *The Spread of Nuclear Weapons.* New York: Norton.

Scharpf, Fritz. 1999. *Governing Europe*. Oxford: Oxford University Press.

Schell, Jonathan. 1982. *The Fate of the Earth*. New York: Knopf.

———. 1984. *The Abolition*. New York: Avon.

———. 2000. The Folly of Arms Control. *Foreign Affairs* 79 no. 5 (September): 22–46.

Schelling, Thomas. 1960. *The Strategy of Conflict*. Cambridge, Mass.: Harvard Univ. Press.

———. 2002. What Makes Greenhouse Sense? *Foreign Affairs* 81 no. 3 (May/June): 2–9.

Schlereth, Thomas. 1977. *The Cosmopolitan Ideal in the Enlightenment*. Notre Dame: The Univ. of Notre Dame Press.

Schmidt, Gavin A. 2007. The Physics of Climate Modeling. *Physics Today* http://www.physicstoday.org/vol-60/iss-1/72_1.html.

Schmitt, Carl. 1927. *The Concept of the Political*. Chicago: Chicago Univ. Press.

———. 1950. *The Nomos of the Earth*. New York: Telos 2003.

Schütz, Alfred, and Thomas Luckmann. 2003. *Strukturen der Lebenswelt*. Konstanz: UVK (English: *The Structure of the Life-World*. Evanston: Northwestern University Press).

Scott, Robert, and Françoise Baylis. 2005. Crossing Species Boundaries. In Shannon, Thomas, ed. *Genetics*. Lanham: Rowman&Littlefield.

Selb, Walter. 1987. *Rechtsordnung und künstliche Reproduktion des Menschen*. Tübingen: Mohr.

Sennett, Richard. 1998. *The Corrosion of Character*. New York: Norton.

Shalit, Avner de. 1995. *Why Posterity Matters*. London: Routledge.

Shrader-Frechette, Karin. 1991. *Risk and Rationality*. Berkeley: Univ. of California Press.

Sikora, Richard, and Brian Barry. 1978. *Obligations to Future Generations*. Philadelphia: Temple Univ. Press.

Singer, Irving. 1992. *Meaning in Life*. New York: The Free Press.

Singer, Pete. 2002. *One World*. New Haven: Yale Univ. Press.

Soroos, Martin. 1997. *The Endangered Atmosphere: Preserving a Global Commons*. Columbia: Univ. of South Carolina Press.

Stephen, Chris, and Allan Hall. 2006. Stalin's half-man, half-ape super-warriors. *The Scotsman*. December 20.

Stern Review. 2006. *The Economics of Climate Change*. http://www.hm-treasury.gov.uk/independent_reviews/stern_review_economics_climate_change/stern_review_report.cfm.

Suganami, Hidemi. 1989. *The Domestic Analogy and World Order Proposals*. Cambridge: Cambridge Univ. Press.

Swazo, Norman. 2002. *Crisis Theory and World Order*. New York: SUNY Press.

Teller, Edward. 1987. *Better a Shield than a Sword*. New York: The Free Press.

Thompson, James. 1986. Psychological Consequences of Disaster: Analogies for the Nuclear Case. In *Institute of Medicine* 1986: 290–314.

Tilly, Charles. 1990. *Coercion, Capital, and European States, AD 990–1990*. Oxford: Blackwell.

Todorov, Tzvetan. 1999. *The Conquest of America: The Question of the Other*. Norman: University of Oklahoma Press.

Toulmin, Stephen. 1990. *Cosmopolis: The Hidden Agenda of Modernity*. New York: Free Press.

———. 2001. *Return to Reason*. Cambridge, Mass.: Harvard Univ. Press.

Transatlantic Trends. 2005. http://www.transatlantictrends.org/trends/doc/2006.

Turner, Frederick J. 1893. The Significance of the Frontier in American History. In Boorstin, Daniel J. *An American Primer*. New York: Penguin 1985, 544–66.

Ungar, Sheldon. 1992. *The Rise and Fall of Nuclearism*. University Park, Penn.: Pennsylvania State University Press.

United Nations (Dept. on Disarmament Affairs). 1989. *Study on the Climatic and Other Global Effects of Nuclear War*. New York.

Urbinati, Nadia. 2003. Can Cosmopolitical Democracy Be Democratic? In Daniele Archibugi, ed. *Debating Cosmopolitics*. London: Verso: 67–85.

Utgoff, Victor A., ed. 2000. *The Coming Crisis: Nuclear Proliferation, U.S. Interests, and World Order*. Cambridge, Mass.: The MIT Press.

Virilio, Paul. 1986. *Speed and Politics: An Essay on Dromology*. New York: Columbia University Press.

Volmerg, Birgit, Ute Volmerg, and Thomas Leithäuser. 1983. *Kriegsängste und Sicherheitsbedürfnisse*. Frankfurt am Main: Fischer.

Von Storch, Hans, and Goetz Flöser. 1999. *Anthropogenic Climate Change*. Berlin: Springer.

Waever, Ole. 2004. *Aberystwyth, Paris, Copenhagen. New 'Schools' in Security Theory and their Origin between Core and Periphery*. http://www.allacademic.com/meta/p74461_index.html.

Waltz, Kenneth. 1954. *Man, the State and War: A Theoretical Analysis*. New York: Columbia University.

Weart, Spencer. 1988. *Nuclear Fear*. Cambridge, Mass.: Harvard Univ. Press.

———. 2003. *The Discovery of Global Warming*. Cambridge, Mass.: Harvard Univ. Press.

Weber, Eugen J. 1999. *Apocalypses*. Cambridge, Mass.: Harvard University Press.

Weber, Max. 1905. *The Protestant Ethic and the Spirit of Capitalism*. London: George Allen & Unwin, 1976.

———. 1906. Critical Studies in the Logic of Cultural Science. In *The Methodology of the Social Sciences*. Glencoe. The Free Press, 1949: 113–88.

———. 1918. The Profession and Vocation of Politics. In *Political Writings*. Cambridge: Cambridge Univ. Press 1994.

———. 1922. *Economy and Society*, vol. 1. Berkeley: Univ. of California Press 1978.

Weber, Steven. 1997. "Institutions and Change." In Doyle, Michael W., and G. John Ikenberry, eds. *New Thinking in International Relations Theory*. Boulder, Colo.: Westview: 229–65.

Weisskopf, Victor F. 1991. *The Joy of Insight*. New York: BasicBooks.

Wendt, Alexander. 1999. *Social Theory of International Politics*. Cambridge: Cambridge Univ. Press.

———. 2003. Why a World State is Inevitable. *European Journal of International Relations* 9 no. 4: 491–542.

Wight, Martin. 1991. *International Theory: the Three Traditions*. Leicester: Leicester Univ. Press.

Winner, Langdon. 1977. *Autonomous Technology*. Cambridge, Mass.: The MIT Press.

World Health Organization. 1984. *Effects of Nuclear War on Health and Health Services*. Geneva.

WorldValuesSurvey. http://www.worldvaluessurvey.org.

Young, Oran R. 2002. *The Institutional Dimension of Environmental Change*. Cambridge, Mass.: The MIT Press.

Index

About the Author

Furio Cerutti is Professor of Political Philosophy at the Department of Philosophy, University of Florence (furio.cerutti@unifi.it). In recent years he has been a Visiting Scholar at the Center for European Studies, Harvard University and Visiting Professor at the Université de Paris 8. A member of GARNET, a Network of excellence funded by the European Union, he coordinates a transnational research group on political identity and legitimacy of the EU and participates in another group on environmental governance.

Among his last publications are two co-edited books: *Political Identities and Conflicts*, Palgrave 2001 and *A Soul for Europe*, two volumes, Peeters, Leuven 2001 (translated into Italian and Farsi).